Applied Mathematical Sciences

EDITORS

Fritz John
Courant Institute of
Mathematical Sciences
New York University
New York, NY 10012

J.E. Marsden
Department of
Mathematics
University of California
Berkeley, CA 94720

Lawrence Sirovich
Division of
Applied Mathematics
Brown University
Providence, RI 02912

ADVISORS

H. Cabannes University of Paris-VI

M. Ghil New York University

J.K. Hale Brown University

J. Keller Stanford University

J.P. LaSalle Brown University

G.B. Whitham California Inst. of Technology

EDITORIAL STATEMENT

The mathematization of all sciences, the fading of traditional scientific bounda-ries, the impact of computer technology, the growing importance of mathematical-computer modelling and the necessity of scientific planning all create the need both in education and research for books that are introductory to and abreast of these developments.

The purpose of this series is to provide such books, suitable for the user of mathematics, the mathematician interested in applications, and the student scientist. In particular, this series will provide an outlet for material less formally presented and more anticipatory of needs than finished texts or monographs, yet of immediate in-terest because of the novelty of its treatment of an application or of mathematics being applied or lying close to applications.

The aim of the series is, through rapid publication in an attractive but inexpen-sive format, to make material of current interest widely accessible. This implies the absence of excessive generality and abstraction, and unrealistic idealization, but with quality of exposition as a goal.

Many of the books will originate out of and will stimulate the development of new undergraduate and graduate courses in the applications of mathematics. Some of the books will present introductions to new areas of research, new applications and act as signposts for new directions in the mathematical sciences. This series will often serve as an intermediate stage of the publication of material which, through exposure here, will be further developed and refined. These will appear in conven-tional format and in hard cover.

MANUSCRIPTS

The Editors welcome all inquiries regarding the submission of manuscripts for the series. Final preparation of all manuscripts will take place in the editorial offices of the series in the Division of Applied Mathematics, Brown University, Providence, Rhode Island.

SPRINGER-VERLAG NEW YORK INC., 175 Fifth Avenue, New York, N.Y. 10010

Printed in U.S.A.

Applied Mathematical Sciences | Volume 45

Applied Mathematical Sciences

(continued)

Klaus Glashoff
Sven-Åke Gustafson

Linear Optimization and Approximation

An Introduction to
the Theoretical Analysis
and Numerical Treatment
of Semi-infinite Programs

With 20 Illustrations

Springer-Verlag
New York Heidelberg Berlin

Klaus Glashoff
Universität Hamburg
Institut für Angewandte
  Mathematik
2 Hamburg 13
Bundestrasse 55
Federal Republic of
  Germany

Sven-Åke Gustafson
Department of Numerical Analysis
  and Computing Sciences
Royal Institute of Technology
S-10044 Stockholm 70
Sweden

and

Centre for Mathematical Analysis
Australian National University
P.O. Box 4
Canberra, ACT 2600
Australia

AMS Subject Classifications: 90C05, 49D35

Library of Congress Cataloging in Publication Data
Glashoff, Klaus, 1947–
    Linear optimization and approximation.
    (Applied mathematical sciences ; v. 45)
    Translation of: Einführung in die lineare Optimierung.
    Includes bibliographical references and index.
    1. Mathematical optimization. 2. Duality theory.
(Mathematics) I. Gustafson, Sven-Ake, 1938–
II. Title. III. Series: Applied mathematical sciences
(Springer-Verlag New York Inc.) ; v. 45.
QA1.A647   vol. 45   510s   [519.7'2]   83-647
[QA402.5]

Original edition © 1978 by Wissenschaftliche Buchgesellschaft, Darmstadt/
West-Germany. (First published in the series: "Die Mathematik. Einführungen in
Gegenstand und Ergebnisse ihrer Teilgebiete und Nachbarwissenschaften.")

English edition © 1983 by Springer-Verlag New York Inc.

Printed and bound by R.R. Donnelley & Sons, Harrisonburg, VA.
Printed in the United States of America.

9 8 7 6 5 4 3 2 1

ISBN 0-387-90857-9 Springer-Verlag New York Heidelberg Berlin
ISBN 3-540-90857-9 Springer-Verlag Berlin Heidelberg New York

Preface

A linear optimization problem is the task of minimizing a linear real-valued function of finitely many variables subject to linear constraints; in general there may be infinitely many constraints. This book is devoted to such problems. Their mathematical properties are investigated and algorithms for their computational solution are presented. Applications are discussed in detail.

Linear optimization problems are encountered in many areas of applications. They have therefore been subject to mathematical analysis for a long time. We mention here only two classical topics from this area: the so-called uniform approximation of functions which was used as a mathematical tool by Chebyshev in 1853 when he set out to design a crane, and the theory of systems of linear inequalities which has already been studied by Fourier in 1823.

We will not treat the historical development of the theory of linear optimization in detail. However, we point out that the decisive breakthrough occurred in the middle of this century. It was urged on by the need to solve complicated decision problems where the optimal deployment of military and civilian resources had to be determined. The availability of electronic computers also played an important role. The principal computational scheme for the solution of linear optimization problems, the simplex algorithm, was established by Dantzig about 1950. In addition, the fundamental theorems on such problems were rapidly developed, based on earlier published results on the properties of systems of linear inequalities.

Since then, the interest of mathematicians and users in linear optimization has been sustained. New classes of practical applications are being introduced continually and special variants of the simplex algorithm and related schemes have been used for the computational treatment of practical problems of ever-growing size and complexity. The theory of "classical" linear optimization problems (with only finitely many linear constraints) had almost reached its final form around 1950; see e.g. the excellent book by A. Charnes, W. W. Cooper and A. Henderson (1953). Simultaneously there were great efforts devoted to the generalization and extension of the theory of linear optimization to new areas. Thus *non-linear* optimization problems were attacked at an early date. (This area plays only a marginal role in our book.) Here, connections were found with the classical theory of Lagrangian multipliers as well as to the duality principles of mechanics. The latter occurred in the framework of convex analysis.

At the same time the theory of *infinite* linear optimization came into being. It describes problems with infinitely many variables and constraints. This theory also found its final form rapidly; see the paper by R. J. Duffin (1956).

A special but important class of infinite linear optimization problems are those problems where the number of variables is finite but the number of linear inequality constraints is arbitrary, i.e. may be infinite. This type of problem, which constitutes a natural generalization of the classical linear optimization problem, appears in the solution of many concrete examples. We have already mentioned the calculation of uniform approximation of functions which plays a major role in the construction of computer representations of mathematical expressions. Uniform approximation can also be successfully used in the numerical treatment of differential equations originating in physics and technological problems.

Using an investigation by Haar from 1924 as a point of departure, A. Charnes, W. W. Cooper and K. O. Kortanek in 1962 gave the fundamental mathematical results of the last-mentioned class of linear optimization problems (with the exception of those questions which were already settled by Duffin's theory).

This class of optimization problems, often called *semi-infinite programs*, will be the main topic of the present book. The "classical" linear optimization problems, called *linear programs*, will occur naturally as a special case.

Whether the number of inequality constraints is finite is a matter of minor importance in the mathematical theory of linear optimization problems. The great advantage of treating such a general class of problems, encompassing so many applications, need not, fortunately, be achieved by means of a correspondingly higher level of mathematical sophistication. In our account we have endeavored to use mathematical tools which are as simple as possible. To understand this book it is only necessary to master the fundamentals of linear algebra and n-dimensional analysis. (This theory is summarized in §2.) Since we have avoided all unnecessary mathematical abstractions, geometrical arguments have been used as much as possible. In this way we have escaped the temptation to complicate simple matters by introducing the heavy apparatus of functional analysis.

The central concept of our book is that of *duality*. Duality theory is not investigated for its own sake but as an effective tool, in particular for the numerical treatment of linear optimization problems.

Therefore all of Chapter II has been devoted to the concept of weak duality. We give some elementary arguments which serve to illustrate the fundamental ideas (primal and dual problems). This should give the reader a feeling for the numerical aspects of duality. In Chapter III we discuss some applications of weak duality to uniform approximation where the emphasis is again placed on numerical aspects.

The duality theory of linear optimization is investigated in Chapter IV. Here we prove theorems on the existence of solutions to the optimization problems considered. We also treat the so-called strong duality, i.e. the question of equality of the values of the primal and dual problems. The "geometric" formulation of the dual problem, introduced here, will be very useful for the presentation of the simplex algorithm which is described in the chapter to follow.

In Chapter V we describe in great detail the principle of the exchange step which is the main building block of the simplex algorithm. Here we dispense with the computational technicalities which dominate many presentations of this scheme. The nature of the simplex algorithm can be explained very clearly using duality theory and the language of matrices and without relying on "simplex tableaux", which do not appear in our text.

In Chapter VI we treat the numerical realization of the simplex algorithm. It requires that a sequence of linear systems of equations be solved. Our presentation includes the stable variants of the simplex method which have been developed during the last decade.

In Chapter VII we present a method for the computational treatment of a general class of linear optimization problems with infinitely many constraints. This scheme was described for the first time in Gustafson (1970). Since then it has been successfully used for the solution of many practical problems, e.g. uniform approximation over multidimensional domains (also with additional linear side-conditions), calculation of quadrature rules, control problems, and so on.

In Chapter VIII we apply the ideas of the preceding three chapters to the special problem of uniform approximation over intervals. The classical Remez algorithm is studied and set into the general framework of linear optimization.

The concluding Chapter IX contains several worked examples designed to elucidate the general approach of this book. We also indicate that the ideas behind the computational schemes described in our book can be applied to an even more general class of problems.

The present text is a translated and extended version of Glashoff-Gustafson (1978). Chapters VIII and IX are completely new and Chapter IV is revised. More material has been added to Chapters III and VII. These changes and additions have been carried out by the second author, who is also responsible for the translation into English. Professor Harry Clarke, Asian Institute of Technology, Bangkok, has given valuable help with the latter task.

We hope that this book will provide theoretical and numerical insights which will help in the solution of practical problems from many disciplines. We also believe that we have clearly demonstrated our conviction that mathematical advances generally are inspired by work on real world problems.

Table of Contents

Chapter I

Introduction and Preliminaries

§1. OPTIMIZATION PROBLEMS

Optimization problems are encountered in many branches of technology, in science, and in economics as well as in our daily life. They appear in so many different shapes that it is useless to attempt a uniform description of them or even try to classify them according to one principle or another. In the present section we will introduce a few general concepts which occur in all optimization problems. Simple examples will elucidate the presentation.

(1) <u>Example</u>: <u>Siting of a power plant</u>. Five major factories are located at $P_1, P_2, \ldots, P_5$. A power plant to supply them with electricity is to be built and the problem is to determine the optimal site for this plant. The transmission of electrical energy is associated with energy losses which are proportional to the amount of transmitted energy and to the distance between power plant and energy consumer. One seeks to select the site of the plant so that the combined energy loss is rendered a minimum. $P_1, P_2, \ldots, P_5$ are represented by points in the plane with the coordinates

$$P_1 = (x_1, y_1), \ldots, P_5 = (x_5, y_5).$$

The distance between the two points $P = (x,y)$, $\bar{P} = (\bar{x}, \bar{y})$ is given by

$$d(P, \bar{P}) = \{(x-\bar{x})^2 + (y-\bar{y})^2\}^{1/2}.$$

Denote the transmitted energy quantities by $E_1, \ldots, E_5$. Our siting problem may now be formulated. We seek, within a given domain G of the plane, a point $\bar{P}$ such that the following function assumes its minimal value at $\bar{P}$:

1

$$E_1 d(P,P_1) + E_2 d(P,P_2) + \ldots + E_5 d(P,P_5).$$

In order to introduce some terminology we reformulate this task. We define the real-valued function f of two real variables x, y through

$$f(x,y) = E_1\{(x-x_1)^2 + (y-y_1)^2\}^{1/2} + \ldots + E_5\{(x-x_5)^2 + (y-y_5)^2\}^{1/2}.$$

We then arrive at the optimization problem: Determine numbers $\bar{x}, \bar{y}$ such that $\bar{P} = (\bar{x},\bar{y}) \in G$ and

$$f(\bar{x},\bar{y}) \leq f(x,y) \quad \text{for all} \quad (x,y) \in G.$$

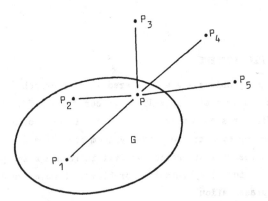

Fig. 1.1.   Siting of power plant

All important concepts associated with optimization problems may be illustrated by this example: f is called a *preference function*, G is the *permissible set*, and the points of G are called *permissible* or *feasible*. Thus the optimization problem means that one should seek a permissible point such that f assumes its minimal value with respect to the permissible set. If such a point does exist, it is called an *optimal point* (for the problem considered), or *optimal solution*, or *minimum point* of f in G.

In the analysis of an optimization problem it is important to verify that an optimal solution does exist, i.e. that the problem is solvable. This is not always the case. As an illustration of this fact we note that the functions $f_1(x) = -x$ and $f_2(x) = e^{-x}$ do not have any minimum points in the set of all real numbers. On the other hand, if an optimization problem is solvable, a minimum point may not be *unique*. In many applications it is required to determine *all* minimum points which the preference function has in the permissible set.

It is of course of no use to formulate a task, appearing in economics or technology, as an optimization problem when this problem cannot
be solved. A formulation as an optimization problem is thus advantageous
only when the mathematical structure of this task can be investigated and
suitable theoretical and computational tools can be brought to bear.
Oftentimes, "applications" to economics or management are proposed
whereby very complicated optimization problems are constructed but it is
not pointed out that neither theoretical nor numerical treatment of the
problem appears to be within reach, now or in the near future. It should
always be remembered that only *some* of the relevant factors can be incor-
porated when a decision problem is formulated as an optimization problem.
There are always decision criteria which cannot be quantified and whose
inclusion into a mathematical model is of doubtful value. Thus, in the
siting problem discussed above, there are many political and ecological
factors which cannot be accounted for in a mathematical model. This indicates that there is, in principle, a limit of what can be gained by the
mathematization of social processes. This difficulty cannot, as a rule,
be overcome by resorting to more complicated models (control theory,
game theory, etc.) even if it sometimes may be concealed. The situation is
quite different for technical systems. Since nowadays the mathematization and also the "optimization" of social processes are pushed forward
with great energy, we find the critical remark above to be justified.

(2) Example: Production model. We consider a firm which produces
or consumes n goods $G_1, \ldots, G_n$ (e.g. raw materials, labor, capital,
environmental pollutants). An *activity* of the firm is represented by n
numbers $(a_1, \ldots, a_n)$ where a_r indicates the amount of good G_r which
is produced or consumed when the activity P is taking place with intensity 1 (measured in suitable units). We assume that the firm can select various activities P_s. Thus the firm's technology has the property
that to each s in a fixed index set S (which may be finite or infinite) there are n numbers $(a_1(s), \ldots, a_n(s))$. A *production plan* of
the firm is defined by selecting a (finite) number of activities
$P_{s_1}, \ldots, P_{s_q}$ and prescribing that they are carried out with the *intensities* $x_1, \ldots, x_q$, where $x_i \geq 0$, i = 1,2,$\ldots$,q. We assume that the production process is linear, i.e. for the given production plan the amount
of good G_r which is produced or consumed is given by

$$a_r(s_1)x_1 + a_r(s_2)x_2 + \ldots + a_r(s_q)x_q.$$

We shall further assume that the activity P_s causes the profit (or cost) $b(s)$. Hence the profit achieved by the chosen production plan is given by

$$b(s_1)x_1 + b(s_2)x_2 + \ldots + b(s_q)x_q. \tag{3}$$

The optimization problem of the firm is to maximize its profit by proper choice of its production plan, i.e. it must select finitely many activities $P_{s_1}, \ldots, P_{s_q}$ and the corresponding intensities $x_1, x_2, \ldots, x_q$ such that the expression (3) assumes the greatest value possible.

The choice of activities and intensities is restricted by the fact that only finite amounts of the goods $G_1, \ldots, G_n$ are available. In practice this is true only for *some* of the goods but for simplicity of presentation we want to assume that *all* goods can only be obtained in limited amounts:

$$a_r(s_1)x_1 + a_r(s_2)x_2 + \ldots + a_r(s_q)x_q \leq c_r, \quad r = 1,2,\ldots,n. \tag{4}$$

Thus (4) defines n *side-conditions* which constrain the feasible activities and intensities. The optimization problem can thus be cast into the form: Determine a finite subset $\{s_1, \ldots, s_q\}$ of the index set S and the real numbers $x_1, \ldots, x_q$ such that the expression (3) is rendered a maximum under the constraints (4) and the further side-conditions

$$x_i \geq 0, \quad i = 1,2,\ldots,q.$$

(5) <u>Remark</u>. A *maximization* problem is transformed into an equivalent *minimization* problem by multiplying its preference function by -1.

(6) <u>The general optimization problem</u>. Let M be a fixed set and let f be a real-valued function defined on M. We seek an element $\bar{x}$ in M such that

$$f(\bar{x}) \leq f(x) \quad \text{for all} \quad x \in M.$$

M is called the *feasible* or *permissible set* and f is termed the *preference function*. We remark here that the feasible set is, as a rule, not explicitly given but is defined through side-conditions (often called *constraints*), as in Example (2).

(7) <u>Definition</u>. The number v given by

$$v = \{\inf f(x) \mid x \in M\}$$

is called the *value* of the corresponding optimization problem. If M is the empty set, i.e. there are no feasible points, the optimization

problem is said to be *inconsistent* and we put $v = \infty$. If feasible points
do exist we term the optimization problem *feasible* or *consistent*. If
$v = -\infty$, the optimization problem is said to be "unbounded from below".
Thus every minimization problem must be in one and only one of the following three "states" IC, B, UB:

IC = Inconsistent; the feasible set is empty and the value of the
problem is $+\infty$.

B = Bounded; there are feasible points and the value is finite.

UB = Unbounded; there are feasible points, the preference function
is unbounded from below, and the value is $-\infty$.

The value of a *maximization problem* is $-\infty$ in the state IC, finite in
state B, and $+\infty$ in the state UB.

§2. SOME MATHEMATICAL PREREQUISITES

The successful study of this book requires knowledge of some elementary concepts of mathematical analysis as well as linear algebra. We
shall summarize the notations and some mathematical tools in this section.

(1) <u>Vectors.</u> We denote the field of real numbers by R, and by R^n
the n-dimensional space of all n-tuples of real numbers

$$\begin{pmatrix} x_1 \\ \vdots \\ x_n \end{pmatrix}. \tag{2}$$

In R^n, the usual vector space operations are defined: componentwise
addition of vectors and multiplication by scalars (i.e. real numbers).

We assume that the reader is familiar with the concepts of "linear
independence", "basis", and "subspace". The zero vector of R^n is
written 0. n-tuples of the form (2) are also referred to as "points".

(3) <u>Matrices.</u> An $m \times n$ matrix A $(m \geq 1)$ is a rectangular array
of $m \cdot n$ real numbers a_{ik} $(i = 1,2,\ldots,m, \ k = 1,2,\ldots,n)$,

$$A = \begin{pmatrix} a_{11} & a_{12} & \cdots & a_{1n} \\ a_{21} & a_{22} & \cdots & a_{2n} \\ \vdots & \vdots & & \vdots \\ a_{m1} & a_{m2} & & a_{mn} \end{pmatrix}.$$

The numbers a_{ik} are termed the *elements* of the matrix A and a_{ik} is situated in row number i and column number k. To each given matrix A we define its *transpose* A^T by

$$A^T = \begin{pmatrix} a_{11} & a_{21} & \cdots & a_{m1} \\ a_{12} & a_{22} & \cdots & a_{m2} \\ \vdots & \vdots & & \vdots \\ a_{1n} & a_{2n} & \cdots & a_{mn} \end{pmatrix}.$$

Every vector $x \in R^n$ may be considered an $n \times 1$ matrix. In order to save space we write, instead of (2),

$$x^T = (x_1, x_2, \ldots, x_n).$$

We note that $(A^T)^T = A$. The reader is supposed to know elementary matrix operations (addition and multiplication of matrices).

(4) <u>Linear mappings</u>. Every $m \times n$ matrix A defines a linear mapping of R^n into R^m whereby every vector $x \in R^n$ is mapped onto a vector $y \in R^m$ via

$$y = Ax. \tag{5}$$

Using the definition of matrix multiplication we find that the components of y are to be calculated according to

$$y_i = a_{i1}x_1 + a_{i2}x_2 + \ldots + a_{in}x_n, \quad 1 \le i \le m.$$

Denote the column vectors of A by $a_1, a_2, \ldots, a_n$. Then we find

$$Ax = a_1 x_1 + a_2 x_2 + \ldots + a_n x_n. \tag{6}$$

Equation (6) thus means that the vector y is a linear combination of the column vectors of A.

(7) <u>Linear systems of equations</u>. Now let a fixed y be given in (5). The task of determining x in (5) is one of the fundamental problems of linear algebra. (5) is called a linear system of equations with n unknowns $x_1, x_2, \ldots, x_n$ and m equations. We assume that the solvability theory of (5) (existence and uniqueness of solutions) is known to the reader. An example: from (6) we conclude that (5) is solvable for each $y \in R^m$ if the column vectors of A span all of R^m, i.e. if A has the *rank* m. It is equally simple to verify that (5) has *at most* one solution if the column vectors of A are linearly independent. The case when A is a *square* matrix, $n \times n$, is of particular interest. Then (5)

has an equal number of equations and unknowns. Then the linear system
$Ax = y$ has a unique solution $x \in R^n$ for each $y \in R^n$ if and only if
the column vectors $a_1, a_2, \ldots, a_n$ of A form a basis of R^n, i.e. are
linearly independent. Then the matrix A is said to be *regular* (or
nonsingular). In this case there exists a $n \times n$ matrix A^{-1} with the
properties

$$A^{-1}(Ax) = x, \quad A(A^{-1}x) = x, \quad \text{all} \quad x \in R^n.$$

A^{-1} is called the *inverse* of A and the linear system of equations (5)
has the unique solution

$$x = A^{-1}y.$$

(8) <u>Hyperplanes</u>. A vector $y \in R^n$ and a number $\eta \in R$ are given.
Then we denote by the *hyperplane* $H(y;\eta)$ the set of all points $x \in R^n$
such that

$$y^T x = y_1 x_1 + y_2 x_2 + \ldots + y_n x_n = \eta.$$

y is called the *normal vector* of the hyperplane. For any two vectors
x and z in $H(y;\eta)$ we have

$$y^T(x-z) = 0.$$

A hyperplane $y^T x = \eta$ partitions R^n into three disjoint sets, namely
$H(y;\eta)$ and the two "open half-spaces"

$$A_1 = \{x \mid y^T x < \eta\}$$
$$A_2 = \{x \mid y^T x > \eta\}.$$

The linear system of equations (5) also admits the interpretation that
the vector x must be in the intersection of the hyperplanes $H(a^i; y_i)$,
$(i = 1, 2, \ldots, m)$, where $a^1, \ldots, a^m$ here are the row-vectors of the matrix
A. Sets of the form $A_1 \cup H(\eta; y)$ and $A_2 \cup H(\eta; y)$ are termed *closed
half-spaces*. They consist of all points $x \in R^n$ such that

$$y^T x \leq \eta \quad \text{or} \quad y^T x \geq \eta,$$

respectively.

(9) <u>Vector norms</u>. We shall associate with each vector $x \in R^n$ a
real number $||x||$. The mapping $x \to ||x||$ shall obey the following
laws:

(i) $||x|| \geq 0$, all $x \in R^n$ and $||x|| = 0$ for $x = 0$ only;

(ii) $||\lambda x|| = |\lambda| \, ||x||$, all $x \in R^n$, all $\lambda \in R$;

(iii) $||x+y|| \leq ||x|| + ||y||$, all $x \in R^n$, $y \in R^n$.

Then $||x||$ will be called the *norm* of the vector.

Exercise: Show that the following mappings define vector norms on R^n:

$$x \to |x_1| + |x_2| + \ldots + |x_n|$$

$$x \to \max\{|x_1|, |x_2|, \ldots, |x_n|\}.$$

The most well-known norm is the Euclidean norm, which will be treated in the next subsection.

(10) Scalar product and Euclidean norm. The *scalar product* of two vectors x and y is defined to be the real number

$$x^T y = y^T x = x_1 y_1 + x_2 y_2 + \ldots + x_n y_n.$$

The real number

$$|x| = \sqrt{x^T x} = (x_1^2 + x_2^2 + \ldots + x_n^2)^{1/2}$$

is called the *Euclidean norm* or *length* or *absolute value* of the vector x. The reader should verify that the mapping $x \to |x|$ defines a norm in the sense of (9). It is also easy to establish the "parallelogram law"

$$|x+y|^2 + |x-y|^2 = 2(|x|^2 + |y|^2) \quad \text{for all} \quad x,y \in R^n.$$

(11) Some topological fundamentals. We define the *distance* between two points x,y in R^n to be given by $|x-y|$. The set $K_r(a)$ consisting of all points whose distance to a is less than r, a fixed positive number, is termed the *open sphere* with *center* a and *radius* r. Thus

$$K_r(a) = \{x \in R^n \mid |x-a| < r\}.$$

We are now in a position to introduce the fundamental topological struc-ture of R^n. A point a is said to be an *inner point* of a subset $A \subset R^n$ if there is a sphere $K_r(a)$ which in its entirety belongs to A, $K_r(a) \subset A$. We will use the symbol $\overset{\circ}{A}$ for the set of all inner points of A. $\overset{\circ}{A}$ is also called the *interior* of A. A is termed *open* if $A = \overset{\circ}{A}$. The point a is said to be a *boundary point* of the set A if every sphere $K_r(a)$ contains both points in A and points which do not belong to A. The set of all boundary points of A is called the *boundary* of A

and is denoted bd A. The union of A and its boundary is called the
closure of A and is denoted $\bar{A}$. The set A is said to be *closed* if
$A = \bar{A}$. The following relations always hold.

$$\overset{\circ}{A} \subset A \subset \bar{A}, \quad \text{bd } A = \bar{A} \smallsetminus \overset{\circ}{A}.$$

The topological concepts introduced above have been defined using the
Euclidean norm. This norm will be most often used in the sequel. However, one may define spheres in terms of other norms and in this way arrive at the fundamental topological concepts "inner points", "open sets",
and so on, in the same manner as above. Fortunately it is possible to
prove that all norms on R^n are *equivalent* in the sense that they generate the same topological structure on R^n: A set which is open with
respect to one norm remains open with respect to all other norms. In order
to establish this assertion one first verifies that if $||\cdot||_1$ and $||\cdot||_2$
are two norms on R^n there are two positive constants c and C such
that

$$c||x||_1 \leq ||x||_2 \leq C||x||_1 \quad \text{for all } x \in R^n.$$

Based on these fundamental structures one can now define the main concept
of convergence of sequences and continuity of functions in the usual way.
We suppose here the reader is familiar with these concepts.

(12) <u>Compact sets</u>. A subset $A \subset R^n$ is said to be *bounded* when
there is a real number $r > 0$ such that $A \subset K_r(0)$. Closed bounded subsets of R^n will be termed *compact*.

Compact subsets A of R^n have the following important property:
Every infinite sequence $\{x_i\}_{i>1}$ of points in the set A has a convergent subsequence $\{x_{i_k}\}_{k>1}$. If $f: R^n \to R^m$ is a continuous mapping, then
the image $f(A)$ of every compact set A is compact also. From this
statement we immediately arrive at the following result which also may be
looked upon as an existence statement for optimization problems:

(13) <u>Theorem of Weierstrass</u>. Let A be a nonempty compact subset
of R^n and f a real-valued continuous function defined on A. Then f
assumes its maximum and minimum value on A, i.e. there exist points
$\bar{x} \in A$ and $\tilde{x} \in A$ such that

$$f(\bar{x}) = \max\{f(x) \mid x \in A\}$$

and

$$f(\tilde{x}) = \min\{f(x) \mid x \in A\}.$$

It is recommended that the reader, as an exercise, carry out the proof of
this simple but important theorem.

§3.   LINEAR OPTIMIZATION PROBLEMS

An optimization problem shall be called a *linear optimization problem*
(LOP) when the preference function is linear and the feasible domain is
defined by linear constraint functions.

Thus the preference function has the form

$$f(y) = c^T y = \sum_{r=1}^{n} c_r y_r,$$

where c is a fixed vector in R^n. The set of feasible vectors of an
(LOP) will be defined as an intersection of half-spaces: Let S be a
given index set (which may be finite or infinite). With each $s \in S$ we
associate a vector $a_s \in R^n$ and a real number b_s. Then the set of
feasible vectors of a linear optimization problem consists of all vectors
$y \in R^n$ lying in *all* half-spaces

$$\{y \mid a_s^T y \geq b_s\}, \quad s \in S. \tag{1}$$

We shall discuss two examples of sets of vectors defined by means of systems of linear inequalities. (In both cases we have $n = 2$.)

(2) **Example**. $S = \{1,2\}$ $a_1 = (2,3)^T$, $a_2 = (-1,0)^T$, $b_1 = 6$, $b_2 = -3$.
In this case (1) becomes

$$2y_1 + 3y_2 \geq 6$$
$$-y_1 \qquad \geq -3.$$

This set is indicated in Figure 3.1 by the checkered area.

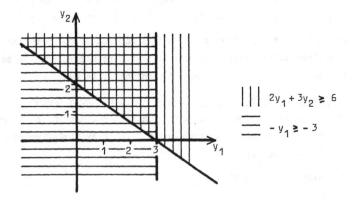

$$|||\quad 2y_1 + 3y_2 \geq 6$$
$$=\quad -y_1 \geq -3$$

Fig. 3.1

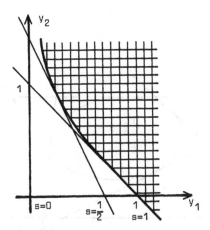

Fig. 3.2. The checkered area is the set defined by means of the inequalities $y_1 + sy_2 \geq \sqrt{s}$, $s \in [0,1]$.

(3) <u>Example</u>. Let S be the real interval $[0,1]$. (S now has infinitely many elements, in contrast to Example (2).) Let $a_s = (1,s)^T$
and $b_s = \sqrt{s}$ for all $s \in [0,1]$. The inequalities (1) then become

$$y_1 + sy_2 \geq \sqrt{s}, \quad s \in [0,1].$$

The subset of the y_1-y_2-plane which is defined by these inequalities is
drawn in Fig. 3.2. The two hyperplanes (in this case straight lines)

$$y_1 + sy_2 = \sqrt{s}$$

corresponding to $s = 1$ and $s = 1/2$ are marked in the figure.

The "general" situation (for $n = 2$) is illustrated in Fig. 3.3.
The hyperplanes corresponding to some particular a_s and b_s, $s \in S$ are
indicated. S may be infinite; if so, it generates infinitely many hyperplanes.

We note that the inequalities (1) may define bounded as well as unbounded subsets of R^n. Compare Fig. 3.2 with Fig. 3.3.

(4) <u>Exercise</u>. Set $n = 2$. Let $S = \{1,2,\ldots,\}$, and let
$a_s = (1,1/s)^T$, $b_s = 0$, for $s = 1,2,\ldots$. Draw the subset of the y_1-y_2-
plane defined by (1). Show that this subset can be defined using two
inequalities only!

(5) <u>Exercise</u>. Draw the subset of the y_1-y_2-plane defined through
the infinitely many inequalities

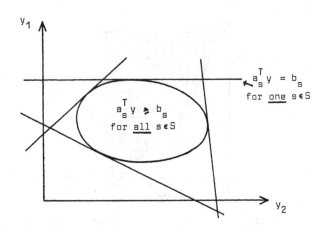

Fig. 3.3

$$-sy_1 - \sqrt{1-s^2}\, y_2 \geq -\sqrt{1-s^2}, \quad s \in [-1,1].$$

To summarize: A *linear optimization problem* is defined as follows:

Given: A vector $c = (c_1, c_2, \ldots, c_n)^T \in R^n$, a nonempty index set S, and for every $s \in S$ a vector $a_s \in R^n$ and a real number b_s.

Sought: A vector $\hat{y} \in R^n$ which solves the following problem (P):

(P)   Minimize $c^T y$ subject to the constraints $a_s^T y \geq b_s$, all $s \in S$.

We now introduce some alternative notations which will often be used in the sequel. We write

　　　　$a(s)$ instead of a_s

and

　　　　$b(s)$ instead of b_s.

Hence we arrive at the following two componentwise representations of the vector $a(s) = a_s$:

　　　　$a_s = (a_{1s}, a_{2s}, \ldots, a_{ns})^T$

and

　　　　$a(s) = (a_1(s), a_2(s), \ldots, a_n(s))^T$.

Thus the optimization problem (P) can also be written in the following form:

(P)   Minimize $\sum_{r=1}^{n} c_r y_r$ subject to the constraints $\sum_{r=1}^{n} a_r(s) y_r \geq b(s)$,

$\quad$ s $\in$ S.

One can use a particularly simple representation in the important special case when S has a *finite* number of elements, i.e. when (P) has only finitely many constraints.

$\quad$ To discuss this case we put $S = \{s_1, s_2, \ldots, s_m\}$ where $m \geq 1$. Then there occur m vectors $a(s_i)$ (i = 1,2,...,m). The corresponding linear constraints take the following form

$$a_1(s_1)y_1 + a_2(s_1)y_2 + \ldots + a_n(s_1)y_n \geq b(s_1)$$
$$a_1(s_2)y_1 + a_2(s_2)y_2 + \ldots + a_n(s_2)y_n \geq b(s_2)$$
$$\vdots \qquad \vdots \qquad \qquad \vdots \qquad \vdots \qquad\qquad (6)$$
$$a_1(s_m)y_1 + a_2(s_m)y_2 + \ldots + a_n(s_m)y_n \geq b(s_m)$$

The nm numbers $a_r(s_i)$ are combined into a matrix A with the vectors $a(s_i)$ in its columns:

$$A = \begin{pmatrix} a_1(s_1) & a_1(s_2) & \cdots & a_1(s_m) \\ a_2(s_1) & a_2(s_2) & \cdots & a_2(s_m) \\ \vdots & \vdots & & \vdots \\ a_n(s_1) & a_n(s_2) & \cdots & a_n(s_m) \end{pmatrix} . \qquad (7)$$

If now the m numbers $b(s_i)$, i = 1,2,...,m are combined into the vector $b = (b(s_1), b(s_2), \ldots, b(s_m))^T$, then the constraints (6) may be written

$$A^T y \geq b.$$

On the other hand let a matrix $A = (a_{rs})$, (r = 1,2,...,n and s = 1,2,...,m) and a vector $b = (b_1, b_2, \ldots, b_m)^T$ be given. Then the inequalities $A^T y \geq b$ become

$$a_{11}y_1 + a_{21}y_2 + \ldots + a_{n1}y_n \geq b_1$$
$$a_{12}y_1 + a_{22}y_2 + \ldots + a_{n2}y_n \geq b_2$$
$$\vdots \qquad \vdots \qquad \qquad \vdots$$
$$a_{1m}y_1 + a_{2m}y_2 + \ldots + a_{nm}y_n \geq b_m$$

This system of inequalities is expressed in the form of (6) by putting

$$S = \{1,2,3,\ldots,m\}$$

and

$$a_r(s) = a_{rs} \quad \text{for} \quad s = 1,2,\ldots,m \quad \text{and} \quad r = 1,2,\ldots,n.$$

(8) __Example__. Consider the system of inequalities

$$y_1 + y_2 \geq 2$$
$$y_1 + 3y_2 \leq 3$$
$$y_1 \geq 0$$
$$y_2 \geq 0.$$

The second inequality is multiplied by -1 and expressed in the form

$$-y_1 - 3y_2 \geq -3.$$

In this case we have $n = 2$, $m = 4$. The matrix A becomes

$$A = \begin{pmatrix} 1 & -1 & 1 & 0 \\ 1 & -3 & 0 & 1 \end{pmatrix}.$$

Every column corresponds to one constraint of the system of inequalities and the corresponding vector b is given by $b = (2,-3,0,0)^T$.

(9) __Definition__. A linear optimization problem with finitely many constraints will be called a _linear program_. Its standard form will be denoted (LP):

(LP)   Minimize $c^T y$ under the constraints $A^T y \geq b$.

Here $A = (a_{rs})$ is a given n by m matrix and b,c are given vectors in R^m and R^n respectively.

   Linear programming, i.e. the algorithmic solution of linear optimization problems of the type (LP), is one of the most important areas of linear optimization. Therefore this special case will be treated separately and in detail in the sequel.

   In the case that (1) defines infinitely many constraints $(|S| = \infty)^*$, it may be advantageous to look upon the vectors a(s) as columns of a "matrix" A. This "matrix" has infinitely many columns. Consider the example of Exercise (4). Here we combine the vectors $a(s) = (1,1/s)^T$ into the array

$$\begin{pmatrix} 1 & 1 & 1 & 1 & \ldots \\ 1 & 1/2 & 1/3 & 1/4 & \ldots \end{pmatrix}.$$

---

*We denote by $|S|$ the number of elements of S. If S has infinitely many elements, we write $|S| = \infty$.

The vectors $a(s)$ can always be arranged in this way when S contains countably many elements but this representation fails in a more general situation, e.g. when $S = [0,1]$. However, also in this case it might be useful to write the vector $a(s)$ from (1) in a matrix-like arrangement. In the case $S = [0,1]$ we may write

$$
\begin{pmatrix}
a_1(0) \ \ldots\ a_1(s) \ \ldots\ a_1(1) \\
a_2(0) \ \ldots\ a_2(s) \ \ldots\ a_2(1) \\
\vdots \qquad \vdots \qquad \vdots \\
a_n(0) \ \ldots\ a_n(s) \ \ldots\ a_n(1)
\end{pmatrix}.
$$
$$
\begin{array}{ccc}
\uparrow & \uparrow & \uparrow \\
a(0) & a(s) & a(1)
\end{array}
$$

(10) <u>Definition</u>. Consider a LOP of the type (P) and such that $|S| = \infty$ (i.e. there are infinitely many linear constraints). Select a *finite subset* $\{s_1, s_2, \ldots, s_m\} \subset S$ and form the matrix A from (7). The linear program hereby arising is called a *discretization* of the original LOP.

As an example we discuss the general LOP:

Minimize $c^T y$ subject to the constraints $\sum_{r=1}^{n} a_r(s) y_r \geq b(s)$, $s \in S$,

where $|S| = \infty$.

A discretization of this task is defined by means of the linear program:

Minimize $c^T y$ subject to the constraints $\sum_{r=1}^{n} a_r(s_i) y_r \geq b(s_i)$,

$i = 1, 2, \ldots, m$.

Here, $s_1, s_2, \ldots, s_m$ are fixed elements in S.

(11) <u>Example</u>. Often problems of the type illustrated by Example (3) are discretized as follows. Select a natural number $m \geq 2$, put $h = 1/(m-1)$, $s_i = (i-1)h$ and form the matrix A. In the case of (3) we get

$$
A = \begin{pmatrix}
1 & 1 & 1 & \ldots & 1 & 1 \\
0 & \dfrac{1}{m-1} & \dfrac{2}{m-1} & \ldots & \dfrac{m-2}{m-1} & 1
\end{pmatrix}.
$$

(12) <u>Exercise</u>. Denote by v the value of Problem (P) and by $v_m(P)$ the value of a discretization of (P). Show that

$$v_m(P) \leq v(P).$$

The method of discretization is very important both in theory and practice. We will return to this topic in §13. Provided that certain very
general conditions are met, it is possible to show that for every linear
optimization problem (P) there is a discretization with the same optimal
solution as (P). These conditions are met in the practical applications
discussed in this book. This statement is an important consequence of the
duality theory of Chapter IV and indicates the important role of linear
programming in the framework of linear optimization.

We mention here that in computational practice discretization is
often used to calculate an approximate solution of a linear optimization
problem with infinitely many constraints. The linear program thereby obtained is solved by means of the simplex algorithm (Chapter V and VI)
which, after finitely many arithmetic operations, delivers a solution
(or the information that none exists).

We shall now illustrate another useful way of studying a given LOP
by means of diagrams.

Consider again Example (3). We have $a(s) = (1,s)^T$, $b(s) = \sqrt{s}$ for
$s \in [0,1]$. Thus

$$a_1(s) = a_{1s} = 1$$
$$a_2(s) = a_{2s} = s$$
$$b(s) = b_s = \sqrt{s}.$$

Let $c_1 = 1$ and $c_2 = 0$. The constraints (1) are written

$$y_1 + sy_2 \geq \sqrt{s}, \quad s \in [0,1].$$

They are illustrated in Fig. 3.2 but may also be represented geometrically
as follows. (y_1,y_2) satisfies these constraints if the straight line

$$z(s) = y_1 + sy_2$$

lies above the graph of the function $\sqrt{s}$ in the interval [0,1]. (See
Fig. 3.4.) The corresponding LOP may be reformulated as the task to
determine, among all such straight lines, the one which intersects the
vertical axis at the lowest point.

(13) **Exercise.** Prove that the LOP above has the *value* 0 but *no*
solution. Show also, by drawing a picture, analogous to Fig. 3.4, that
every discretization of this LOP has the value $-\infty$, if the left boundary
point of the interval [0,1] does *not* appear among the points of discretization, $s_1, s_2, \ldots, s_m$. Thus the linear program is unbounded from below in this case.

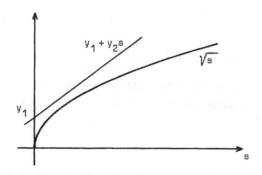

Fig. 3.4

(14) **Example: Air pollution control.** We consider the problem of maintaining a satisfactory air quality in an area S (e.g. a city). This goal shall be reached by regulating the emissions from the sources of pollutants in such a manner that the control costs are as small as possible. N sources have been identified and their positions and strengths are known.

We consider here only the case of *one* pollutant, e.g. SO_2. The concentration of the pollutant at a point $s = (s_1, s_2)^T$ is given by

$$d(s) = \sum_{j=1}^{N} q_j V_j(s).$$

Here V_j is the *transfer function* which describes the contribution from the source with index j to the ambient concentration at the point s. V_j describes an *annual* mean and is hence time-independent. The transfer functions are calculated from meteorological dispersion models incorporating wind speed and direction, atmospheric stability, and several other geographical and meteorological variables. We shall assume that the transfer functions are known. q_j is the strength of source number j.

The number of pollutant sources is generally very great and therefore they cannot be regulated individually. Instead they are divided into n source classes $G_1, G_2, \ldots, G_n$ and all sources in a given class are regulated in the same way. Thus all residential houses of a city may form one source class. The sources are now numbered so that all sources with indices between $j_{r-1} + 1$ and j_r comprise class number r $(r = 1, 2, \ldots, n)$.

Thus we have

$$0 = j_0 < j_1 < \ldots < j_n = N.$$

We now introduce

$$v_r(s) = \Sigma \, q_j V_j(s) \qquad (r = 1,2,\ldots,n)$$

where the summation is extended over all members of class r. The total concentration of the pollutant at point s is thus given by

$$\sum_{r=1}^{n} v_r(s).$$

One reduction strategy is now to reduce the emmission of class G_r by the fraction E_r. Thus $0 \le E_r \le 1$ $(r = 1,2,\ldots,n)$. Hence the total remaining concentration after regulation becomes

$$\sum_{r=1}^{n} (1 - E_r)v_r(s).$$

We require now that for each $s \in S$ the value of this expression does not surpass a given limit g(s). g may be a legally imposed standard defining the highest acceptable concentration. We assume also that there are upper bounds $e_r < 1$ for the fractions E_r. (It is not technically possible to completely remove the emissions from the group G_r.) Therefore the numbers $E_1, E_2, \ldots, E_n$ must meet the conditions:

$$0 \le E_r \le e_r, \qquad r = 1,2,\ldots,n \tag{15}$$

$$\sum_{r=1}^{n} (1-E_r)v_r(s) \le g(s), \quad s \in S. \tag{16}$$

The reduction of emissions entails costs, e.g. for the installment and maintenance of effluent filters in factories. We shall assume that these costs are defined by the linear function

$$K(E) = \sum_{r=1}^{n} c_r E_r, \tag{17}$$

where $c_1, c_2, \ldots, c_n$ are known numbers. The task of minimizing the cost function (17) under the constraints (15), (16) is a linear optimization problem:

Minimize $\displaystyle\sum_{r=1}^{n} c_r E_r$ subject to the constraints $E_r \ge 0$, $r = 1,2,\ldots,n,$

$$-E_r \geq -e_r, \quad r = 1,2,\ldots,n,$$

$$\sum_{r=1}^{n} E_r v_r(s) \geq -g(s) + \sum_{r=1}^{n} v_r(s), \quad s \in S.$$

· (18) **Remark**. The function d does not completely describe the air quality since the level of concentration changes irregularly with time. The reduction policy which is determined by considering the annual mean concentrations only is therefore a long-term regulation strategy which must be supplemented with suitable short-term measures to counteract temporary strong increases in ambient concentrations.

The above formulation of an optimization problem for environmental pollution control is based on work by Gorr and Kortanek. See e.g. Gorr, Gustafson and Kortanek (1972) and Gustafson and Kortanek (1975).

Chapter II
Weak Duality

The present chapter is very elementary in its entirety but is of decisive importance for understanding the material to follow. Here we lay the foundations for the theoretical as well as computational treatment of linear optimization problems. The simple examples are particularly designed in order to familiarize the reader with the structure of such problems as well as the central concept of duality which plays a major role both in the theory and in all practical applications of linear optimization. A thorough study of these examples is the best preparation for the duality theory to be presented in Chapter IV and the algorithms of Chapters V through VIII.

§4. DUALITY LEMMA AND DUAL PROBLEM

We consider the optimization problem (P) which was introduced in §3. It can be written in the following compact form:

(P)   Minimize $c^T y$ subject to $a(s)^T y \geq b(s)$, $s \in S$

or alternatively

(P)   Minimize $\sum_{r=1}^{n} c_r y_r$ subject to $\sum_{r=1}^{n} a_r(s) y_r \geq b(s)$, $s \in S$.

One obtains an upper bound for the value $v(P)$ as soon as a feasible vector y is available. According to the definition of $v(P)$ we find immediately that

$$v(P) \leq c^T y.$$

It is of great interest for numerical treatment to determine good *lower*
bounds for $v(P)$. This fact will be illustrated in many examples. The
following fundamental lemma can be used for constructing such lower bounds.

(1) **Duality lemma.** Let the finite subset $\{s_1, s_2, \ldots, s_q\} \subset S$,
$q \geq 1$, and the nonnegative numbers $x_1, x_2, \ldots, x_q$ be such that

$$c = a(s_1)x_1 + a(s_2)x_2 + \ldots + a(s_q)x_q. \tag{2}$$

Then the following inequality holds for every feasible vector
$y = (y_1, \ldots, y_n)^T$:

$$b(s_1)x_1 + b(s_2)x_2 + \ldots + b(s_q)x_q \leq c^T y. \tag{3}$$

Proof: We have assumed that y is feasible for (P). Then we find
in particular

$$a(s_i)^T y \geq b(s_i), \quad i = 1, 2, \ldots, q.$$

Since $x_i \geq 0$, $i = 1, 2, \ldots, q$, we get

$$\sum_{i=1}^{q} b(s_i)x_i \leq \sum_{i=1}^{q} (a(s_i)^T y)x_i = \left(\sum_{i=1}^{q} a(s_i)x_i \right)^T y.$$

The assertion now follows from (2).

Since (3) holds for *every* vector y which is feasible for (P) we
immediately arrive at the following statement on lower bounds for the
optimal value $v(P)$. (Note that here we revert to the componentwise
representation of the vectors $a(s_i)$ and c.)

(4) **Corollary.** Let $\{s_1, \ldots, s_q\}$, $q \geq 1$ be a finite subset of the
index set S and let the numbers $x_1, \ldots, x_q$ satisfy

$$\sum_{i=1}^{q} a_r(s_i)x_i = c_r, \quad r = 1, 2, \ldots, n. \tag{5}$$

Then

$$\sum_{i=1}^{q} b(s_i)x_i \leq v(P). \tag{6}$$

We remark already here that one is, of course, interested in obtaining
the best possible lower bounds for $v(P)$. We will show in later chapters
that for large classes of problems it is possible to obtain *arbitrarily*
good lower bounds by selecting the subset $s_1, \ldots, s_q$ and the numbers x_i
properly.

(7) **Example.** We consider the LOP

Minimize $y_1 + 1/2 \, y_2$ subject to $y_1 + sy_2 \geq e^s$, $s \in [0,1]$.

We try now to determine a finite subset $\{s_1,\ldots,s_q\}$ of S and nonnegative numbers $x_1,\ldots,x_q$ such that the assumptions of the duality lemma
are met. We take first $q = 1$ and seek a point s_1 in the interval
[0,1] and a nonnegative number x_1 with the property (5):

$$1 \cdot x_1 = 1$$
$$s_1 \cdot x_1 = 1/2.$$

These equations have the unique solution $x_1 = 1$, $s_1 = 1/2$. From (6) we
get

$$x_1 e^{s_1} = 1 \cdot e^{1/2} = \sqrt{e} \cong 1.648 \ldots \leq v(P).$$

It is also easy to obtain a rough *upper* bound: One needs only to find
numbers $\tilde{y}_1, \tilde{y}_2$ such that the straight line $\tilde{y}_1 + s\tilde{y}_2$ lies above the
curve e^s throughout the interval [0,1]. This occurs e.g. for
$\tilde{y}_1 = 1$, $\tilde{y}_2 = 2$. (Draw a picture similar to Fig. 3.4.) We get $v(P) \leq$
$\tilde{y}_1 + 1/2 \, \tilde{y}_2 = 2$. Hence we have arrived at the (not very good) bracketing

$$1.648 \leq v(P) \leq 2.$$

A better result is obtained by selecting $q = 2$. We then are faced with
the equations (see (5)):

$$x_1 + x_2 = 1$$
$$s_1 x_1 + s_2 x_2 = 1/2.$$

One possible solution is given by $s_1 = 0$, $s_2 = 1$, $x_1 = x_2 = 1/2$. From
(6),

$$x_1 e^{s_1} + x_2 e^{s_2} = 1/2 + 1/2 \cdot e \cong 1.859 \leq v(P).$$

(8) **Exercise.** Show that indeed,

$$v(P) = 1/2(1+e)$$

by determining a suitable upper bound.

(9) **Example.** Consider the linear program

Minimize $3y_1 + y_2$

subject to the constraints of Example (8) in §3. We seek a lower bound
for its optimal value. To obtain a representation (2) or (5) means that

the vector $c = (3,1)^T$ shall be written as a nonnegative linear combination of q columns of the matrix appearing in Example (8) in §3:

$$A = \begin{pmatrix} 1 & -1 & 1 & 0 \\ 1 & -3 & 0 & 1 \end{pmatrix}.$$

Since $c \in R^2$, we take $q = 2$ and try at first to represent c as a nonnegative linear combination of the first columns of A.

$$\begin{pmatrix} 1 \\ 1 \end{pmatrix} x_1 + \begin{pmatrix} -1 \\ -3 \end{pmatrix} x_2 = \begin{pmatrix} 3 \\ 1 \end{pmatrix}.$$

The unique solution of this linear system of equations turns out to be $x_1 = 4$, $x_2 = 1$. From (6) we now get the lower bound 5 for the optimal value. (We had $b = (2,-3,0,0)^T$.) Determine graphically the optimal value and the solution of the linear program.

(10) **Lemma.** Let $y = (y_1, \ldots, y_n)$ be feasible for the problem (P). Assume also that the subset $\{s_1, \ldots, s_q\}$ of S and the nonnegative numbers $x_1, \ldots, x_q$ satisfy the assumption (2) of the duality lemma. If

$$\sum_{i=1}^{q} b(s_i) x_i = \sum_{r=1}^{n} c_r y_r \tag{11}$$

is satisfied, then y is an optimal solution to (P).

Proof: Since y is feasible for (P) we have

$$v(P) \leq \sum_{r=1}^{n} c_r y_r.$$

On the other hand, from (11) and (6),

$$\sum_{r=1}^{n} c_r y_r \leq v(P).$$

The assertion follows.

(12) **Linear programming.** Consider now the particular problem

(LP)   Minimize $c^T y$ subject to $A^T y \geq b$,

where A has m column vectors $a_1, \ldots, a_m$. In this case $q \leq m$ must hold, of course. Then every nonnegative solution of the system

$$Ax = c, \quad x = (x_1, \ldots, x_m)^T \tag{13}$$

will give lower bounds for the value $v(LP)$ of the form

$$b^T x \leq v(LP). \tag{14}$$

Note that (13) can be written in the alternative form

$$c = \sum_{i=1}^{m} a_i x_i,$$

which corresponds to Equation (5), while (14) corresponds to the inequality (6).

A natural objective is to select the subset $\{s_1, \ldots, s_q\}$ and the nonnegative numbers $x_1, \ldots, x_q$ in order to maximize the lower bound for the value $v(LP)$ obtained from the duality lemma. We arrive at the

Dual problem (D): Determine a finite subset $\{s_1, \ldots, s_q\} \subset S$ and real numbers $x_1, \ldots, x_q$ such that the expression

$$\sum_{i=1}^{q} x_i b(s_i) \tag{15}$$

is maximized, subject to the constraints

$$\sum_{i=1}^{q} x_i a_r(s_i) = c_r, \qquad r = 1, 2, \ldots, n, \tag{16}$$

$$x_i \geq 0, \quad i = 1, 2, \ldots, q. \tag{17}$$

$\{s_1, \ldots, s_q, x_1, \ldots, x_q\}$ is said to be *feasible for (D)* when $s_i \in S$, $i = 1, 2, \ldots, q$, and (16) and (17) hold.

The problem (D) appears to be very complicated since q, the number of "mass points", may be arbitrarily large. However, we will see in Chapter IV that $q = n$ may be assumed in all problems of practical interest. (Then (D) is a nonlinear optimization problem with $2n$ variables.) But in our argument we shall start by allowing q to be arbitrarily large. Denote by $v(D)$ the value of (15) subject to (16) and (17). Then we conclude from the duality lemma (1) the

(18) Weak duality theorem. $v(D) \leq v(P)$

The pair of problems (P) - (D) is called a *dual pair*. The transfer from the primal problem (P) to the dual problem (D) will be called *dualization*. The following reformulation of Lemma (10) will be useful when the results of the present section are applied to concrete problems.

(19) Lemma. Let $y = (y_1, \ldots, y_n)^T$ be feasible for (P) and $\{s_1, \ldots, s_q, x_1, \ldots, x_q\}$ be feasible for (D). If

$$\sum_{i=1}^{q} b(s_i)x_i = \sum_{r=1}^{n} c_r y_r$$

holds, then y is a solution of (P) and $\{s_1,\ldots,s_q,\ x_1,\ldots,x_q\}$ is a solution of (D).

(20) <u>Complementary slackness lemma</u>. Let $y = (y_1,\ldots,y_n)^T$ be feasible for (P) and $\{s_1,\ldots,s_q,\ x_1,\ldots,x_q\}$ be feasible for (D). Assume also that the following relation holds:

$$x_i\left(\sum_{r=1}^{n} a_r(s_i)y_r - b(s_i)\right) = 0, \quad i = 1,\ldots,q . \tag{21}$$

Then y is a solution of (P) and $\{s_1,\ldots,s_q,\ x_1,\ldots,x_q\}$ is a solution of (D). Further, the values of (P) and (D) coincide.

<u>Proof</u>: In (21), $x_i > 0$ implies

$$\sum_{r=1}^{n} a_r(s_i)y_r = b(s_i), \quad i = 1,2,\ldots,q.$$

Thus we have the following equation:

$$\sum_{i=1}^{q} b(s_i)x_i = \sum_{i=1}^{q}\left(\sum_{r=1}^{n} a_r(s_i)y_r\right)x_i = \sum_{r=1}^{n}\left(\sum_{i=1}^{q} a_r(s_i)x_i\right)y_r = \sum_{r=1}^{n} c_r y_r.$$

Here we have used the feasibility of $\{s_1,\ldots,s_q,\ x_1,\ldots,x_q\}$. The assertion now follows from Lemma (19).

(22) <u>Example</u>: <u>Optimal production plan</u>. In this subsection we return to the production model (2) in §1. There we considered n *goods* $G_1,\ldots,G_n$ and the possible *activities* P_s ($s \in S$) which were described by the vectors

$$a(s) = (a_1(s),\ldots,a_n(s))^T.$$

Here $a_r(s)$ is a measure of the amount of good G_r which is consumed or produced when activity P_s is carried out with intensity 1.

We had formulated an optimization problem (for maximization of profits) of the following form: Determine a finite subset $\{s_1,\ldots,s_q\}$ ($q \geq 1$) of the index set S and real numbers $\{x_1,\ldots,x_q\}$ such that the expression

$$b(s_1)x_1 + b(s_2)x_2 + \ldots + b(s_q)x_q \tag{23}$$

is maximized subject to the constraints

$$a_r(s_1)x_1 + a_r(s_2)x_2 + \ldots + a_r(s_q)x_q \leq c_r, \quad r = 1,\ldots,n, \tag{24}$$

and

$$x_i \geq 0, \quad i = 1,\ldots,q. \tag{25}$$

In order to get an optimization problem of the type (D) we introduce slack variables ξ_r, $r = 1,2,\ldots,n$. Then we write (24) - (25) in the following equivalent form

$$\sum_{i=1}^{q} a_r(s_i)x_i + \xi_r = c_r, \quad r = 1,2,\ldots,n \tag{26}$$

$$x_i \geq 0 \quad (i = 1,\ldots,q), \quad \xi_r \geq 0 \quad (r = 1,\ldots,n). \tag{27}$$

This may be interpreted as meaning that the activities P_s, $s \in S$ are supplemented with the so-called *disposal-activities* P_r, $r = 1,\ldots,n$.

(28) The corresponding primal problem. The maximization of the pre-ference function (23) subject to the constraints (26), (27) is the dual of the following linear optimization problem:

$$\text{Minimize} \quad \sum_{r=1}^{n} y_r c_r \tag{29}$$

subject to the constraints

$$\sum_{r=1}^{n} a_r(s)y_r \geq b(s), \quad s \in S \tag{30}$$

$$y_r \geq 0, \quad r = 1,\ldots,n. \tag{31}$$

The variables $y_1,\ldots,y_n$ of this primal problem may be interpreted as the *prices* of the goods $G_1,\ldots,G_n$ and the number

$$\sum_{r=1}^{n} a_r(s)y_r \tag{32}$$

indicates the *cost* which arises when the activity P_s ($s \in S$) is carried out with intensity 1. Thus a "price system" $y_1,\ldots,y_n$ is feasible (i.e. meets the conditions (30) - (31)) when all prices are nonnegative and when the cost (32) for no $s \in S$ is below the revenue $b(s)$ result-ing when the activity P_s is carried out with unit intensity. The com-plementary slackness lemma (20) now assumes the following form:

(33) Let $\{s_1,\ldots,s_q, x_1,\ldots,x_q\}$ be a feasible production plan with $x_i > 0$ for $i = 1,\ldots,q$ and let y be a feasible price vector. These production plans and price vectors are optimal if

$$\sum_{r=1}^{n} a_r(s_i)y_r = b(s_i), \quad i = 1,\ldots,q\,, \tag{34}$$

and

$$y_r\xi_r = 0, \quad r = 1,\ldots,n, \tag{35}$$

with

$$\xi_r = c_r - \sum_{i=1}^{q} a_r(s_i)x_i, \quad r = 1,\ldots,n.$$

The conditions (34) and (35) admit an excellent economic interpretation:
A feasible production plan and a feasible price vector are optimal if
i) the cost per unit intensity of each activity P_s occurring in the production plan is equal to the corresponding revenue $b(s)$ and if ii) the
prices y_r of goods G_r which are not exhausted (i.e. $\xi_r > 0$) are zero.

By means of the tools developed in Chapter IV we will be able to
give conditions which ensure that the problem (23) - (25) of finding an
optimal production plan is solvable. We shall also demonstrate that
there is then an optimal production plan involving at most n activities.
This result is true even if there are arbitrarily many possible activities.

The study of production models of the same kind as, and similar to,
that of problem (23) - (25) has greatly stimulated the development of
linear programming. The whole theory of Chapter IV as well as the simplex
algorithm of Chapter V can be motivated with concepts from economics.
This is expounded in the book by Hildenbrand and Hildenbrand (1975) and
the reader is referred to this text.

(36) Duality for linear programming. We now investigate the important special case of *linear programming*, i.e. when the index set S is
finite, $S = \{1,\ldots,m\}$. Then (P) takes the special form (see (9), §3):

(LP)   Minimize $\sum_{r=1}^{n} c_r y_r$ subject to $A^T y \geq b.$

We recall that the constraints of (LP) may be written in the form

$$a_j^T y \geq b_j, \quad j = 1,\ldots,m,$$

where $a_1,\ldots,a_m$ are the column vectors of the matrix A, and

$$A = \begin{pmatrix} a_{11} & a_{12} & \cdots & a_{1m} \\ a_{21} & a_{22} & \cdots & a_{2m} \\ \vdots & \vdots & & \vdots \\ a_{n1} & a_{n2} & \cdots & a_{nm} \\ \uparrow & \uparrow & & \uparrow \\ a_1 & a_2 & & a_m \end{pmatrix}, \qquad b = \begin{pmatrix} b_1 \\ b_2 \\ \vdots \\ b_m \end{pmatrix}.$$

In this case there are only finitely many vectors a_i $(i = 1,\ldots,m)$ and $x_i = 0$ is permitted by the constraints of the dual problem. Therefore we may put $q = m$ from the outset and replace (16), (17) by

$$\sum_{i=1}^{m} a_i x_i = c, \quad x_i \geq 0 \quad \text{for} \quad i = 1,\ldots,m.$$

Using matrices we get with $x = (x_1,\ldots,x_m)^T$

$$Ax = c, \quad x \geq 0.$$

Therefore we define the *dual linear program* to be the optimization problem

(LD)   Maximize $\displaystyle\sum_{i=1}^{m} b_i x_i = b^T x$ subject to $Ax = c, \quad x \geq 0.$

This is a problem with a linear preference function, linear equality constraints, and positivity requirements for all variables. It is a very important fact that problems of the type (LP) through simple transformations can be brought into the form (LD) and vice versa. This is not possible for general problems of the type (P) and (D).

(37) <u>The transformation (LP) $\to$ ($\hat{\text{LD}}$)</u>. A vector $y \in R^n$ meets the constraints

$$A^T y \geq b$$

of (LP) if and only if there is a vector $z \in R^m$ such that

$$A^T y - z = b, \quad z \geq 0 \tag{38}$$

(Such a z is called a *slack* vector). This system of equations and inequalities to be satisfied by the vector $(y,z) \in R^{n+m}$ does not have the same form as the constraints of (LD) since only some of the $n+m$ variables, namely $z_1,\ldots,z_m$, must be nonnegative. This is remedied by splitting up y in the following way. Consider the system

$$A^T y^+ - A^T y^- - z = b$$
$$y^+ \geq 0, \quad y^- \geq 0, \quad z \geq 0 \tag{39}$$

where $y^+ \in R^n$, $y^- \in R^n$, $z \in R^m$. We show that (39) and (38) are equivalent. If y^+, y^- and z satisfy (39), then the vectors $y = y^+ - y^-$
and z satisfy (38). To prove the converse note that every vector
$y \in R^n$ may be written

$$y = y^+ - y^- \quad \text{with} \quad y^+ \geq 0, \quad y^- \geq 0. \tag{40}$$

Thus from any solution (y,z) of (38) we may construct a solution
(y^+, y^-, z) of (39). A representation (40) of y may be obtained by
putting

$$\left. \begin{array}{l} y_r^+ = \max(y_r, 0) \\ y_r^- = -\min(y_r, 0) \end{array} \right\}, \quad r = 1, \ldots, n. \tag{41}$$

But the representation $y = y^+ - y^-$ is not the only possible one of the
type (40). Let

$$\left. \begin{array}{l} \hat{y}_r^+ = y_r^+ + \alpha_r \\ \hat{y}_r^- = y_r^- + \alpha_r \end{array} \right\}, \quad r = 1, \ldots, n, \tag{42}$$

where α_r are arbitrary nonnegative numbers. Then $y = \hat{y}^+ - \hat{y}^-$ is also
a representation of the type (40) and it is easy to show that all representations of the type (40) may be constructed from (42). We observe now
that

$$c^T y = c^T \hat{y}^+ - c^T \hat{y}^-$$

holds for all representations of the type (42). Therefore it follows that
the program (LP) is equivalent to the following optimization problem of
type (LD):

Maximize $-(c^T y^+ - c^T y^-)$ subject to

$$(\hat{LD}) \quad (A^T, \quad -A^T, \quad -I_m) \begin{pmatrix} y^+ \\ y^- \\ z \end{pmatrix} = b$$

$$(y^+, \quad y^-, \quad z)^T \geq 0.$$

(43) <u>The transformation (LD) $\rightarrow$ ($\hat{LP}$)</u>. We rewrite the constraints of
(LD),

$$Ax = c, \quad x \geq 0,$$

in the equivalent form

$$Ax \geq c$$
$$-Ax \geq -c$$
$$x \geq 0.$$

Then we obtain from (LD) the following optimization problem of type (LP):

$$(\hat{LP}) \quad \text{Minimize } -b^T x \text{ subject to } \begin{pmatrix} A \\ -A \\ I_m \end{pmatrix} x \geq \begin{pmatrix} c \\ -c \\ 0 \end{pmatrix}.$$

(44) **Definition.** We define the *double dualization* of the linear program (LP) to be the following process: First the linear program (LP) is dualized giving (LD). Then the transformation (43) (LD) → $(\hat{LP})$ is carried out. Lastly, the linear program $(\hat{LP})$ is dualized.

We see immediately that $(\hat{LD})$ is the dual of $(\hat{LP})$. But we have already shown that (LP) and $(\hat{LD})$ are equivalent. Thus we arrive at the important result:

(45) **Theorem.** If the linear program (LP) undergoes a double dualization, an optimization problem equivalent to (LP) results.

(46) **Exercise.** Consider the two optimization problems

$$\text{Minimize } c^T y \text{ subject to } Ay \geq b, \quad y \geq b, \quad y \in R^n,$$

and

$$\text{Maximize } b^T x \text{ subject to } A^T x \leq c, \quad x \geq 0, \quad x \in R^m.$$

In what sense can they be said to form a dual pair? Carry out suitable transformations which bring them into the form (LP) or (LD).

§5. STATE DIAGRAMS AND DUALITY GAPS

Using the simple weak duality theorem (18) of §4, we may immediately derive a first classification table for the dual pair (P) - (D). (Results of the type $v(P) = v(D)$ are called *strong* duality theorems. They are given in Chapter IV.) We recall that every minimization problem of the type (P) must be in one and only one of the three states (see (7), §1)

      IC (Inconsistent; there are no feasible vectors y. By definition we have $v(P) = \infty$.)

(P)   B   (Bounded; there are feasible vectors y and $v(P)$ is finite.)

      UB (Unbounded; there are feasible vectors y such that the preference function is arbitrarily small, i.e. $v(P) = -\infty$.)

By the same token, the dual problem must be in one and only one of the
three states indicated below. (Observe that (D) is a *maximization* problem.)

      IC   (Inconsistent: $v(D) = -\infty$.)

(D)   B   (Bounded: $v(D)$ finite.)

      UB (Unbounded: $v(D) = +\infty$.)

The statement of the duality theorem (18) of §4 may be represented by the
state diagram below. Combinations of states of the dual pair (P) - (D)
which are impossible by (18) of §4 are marked with a cross in the diagram.
(The reader should verify that these combinations cannot occur.)

(1) <u>State diagram for the dual pair (P) - (D)</u>.

D \ P	IC	B	UB
IC	1	2	4
B	3	5	x
UB	6	x	x

The Case 5 is of main interest for the applications. Then (P) and (D)
are both bounded. This occurs when both problems are feasible.

    It is possible to construct simple examples to demonstrate that all
the Cases 1,2,3,4,5, and 6, which are not excluded by the weak duality
theorem, do in fact occur in practice.

    We will show later that the Cases 2 and 3 do not occur in linear
programming, i.e. linear optimization problems of type (LP). It is often
possible to introduce "reasonable" assumptions on general linear optimization problems in order to insure that Cases 2 and 3 do not materialize.
We shall treat this topic in detail in Chapter IV. Nevertheless, we illustrate Cases 2 and 3 of the state diagram by means of two examples constructed for the purpose.

(2) <u>Example</u>. $n = 2$, $S = [0,1]$.

(P)   Minimize y_1 subject to the constraints $sy_1 + s^2 y_2 \geq s^2$, $s \in S$.

(P) has feasible vectors, for we may take $y_1 = 0$, $y_2 = 1$. Furthermore,
all feasible vectors $y = (y_1, y_2)^T$ must satisfy $y_1 \geq 0$. This fact is
easily illustrated by means of a diagram similar to Fig. 3.4. Therefore
we get

    $v(P) = 0$

and Problem (P) is hence in State B.

The corresponding dual problem (D) reads

$$\text{Maximize} \quad \sum_{i=1}^{q} s_i^2 x_i$$

subject to the constraints

$$\sum_{i=1}^{q} s_i x_i = 1 \tag{3}$$

$$\sum_{i=1}^{q} s_i^2 x_i = 0 \tag{4}$$

$$\left.\begin{array}{l} s_i \in [0,1] \\[1.5ex] x_i \geq 0 \end{array}\right\} \quad \text{for} \quad i = 1,\ldots,q \quad \text{and} \quad q \geq 1.$$

The *inconsistency* of (D) is shown as follows: By (4), for $i = 1,\ldots,q$ we must have $x_i = 0$ or $s_i = 0$ since $x_i \geq 0$ and $s_i^2 \geq 0$. But then (3) cannot be satisfied. (D) is therefore in State IC and we have thus an instance of Case 2 in diagram (1).

(5) <u>Example</u>. $n = 1$, $S = [0,1]$

(P)   Minimize $0 \cdot y_1$ subject to the constraints $s^2 y_1 \geq s$, $s \in S$.

Since $s^2 y_1 \geq s \longleftrightarrow s(sy - 1) \geq 0$, each feasible y_1 must satisfy $sy_1 - 1 \geq 0$ for all $s \in [0,1]$. This is not possible for any number y_1, implying that (P) is in State IC.

The dual problem is

(D)
$$\text{Maximize} \quad \sum_{i=1}^{q} s_i x_i \quad \text{subject to the constraints} \quad \sum_{i=1}^{q} s_i^2 x_i = 0,$$

$$s_i \in [0,1], \quad x_i \geq 0, \quad \text{for} \quad i = 1,\ldots,q \quad (q \geq 1),$$

(D) is feasible and for each permissible $\{s_1,\ldots,s_q,\ x_1,\ldots,x_q\}$ it follows that $s_i = 0$ or $x_i = 0$ for $i = 1,\ldots,q$. Thus (D) is in State B, hence we have an instance of Case 3 in diagram (1).

We have already mentioned that we shall in Chapter IV establish theorems proving $v(P) = v(D)$ is true given certain general assumptions. Thus we will prove that $v(LP) = v(LD)$ always holds for linear programming if at least one of the problems is feasible. However, at the end of this section we shall give examples of linear optimization problems which are in Case 5 of the diagram (1); i.e. where both the primal and dual problems are bounded, but where $v(P)$ and $v(D)$ do not coincide.

(6) <u>Definition</u>. Let a dual pair (P) - (D) be given. The number

$$\delta(P,D) = v(P) - v(D)$$

is called the *defect*. We introduce here the convention

$+\infty - (-\infty) = +\infty$

$+\infty - (+\infty) = 0$

$-\infty - (-\infty) = 0$

$c - (-\infty) = +\infty$

$+\infty - c = +\infty$

for all real numbers c. If $\delta(P,D) > 0$, we say that a *duality gap* has occurred.

The following diagram gives the values of the defect corresponding to all states of the dual pair. This diagram is obtained directly from the state diagram (1). (The impossible states which are marked with a cross in (1) are omitted.)

(7) <u>Defect diagram</u>.

(D) \ (P)	IC	B	UB
IC	$+\infty$	$+\infty$	0
B	$+\infty$	d	
UB	0		

Here d stands for a nonnegative number, $0 \le d \le \infty$.

(8) <u>Example</u>. Consider the following problem of type (P):

Minimize y_1 subject to the constraints $sy_1 + s^2 y_2 \ge 0$, $s \in [0,1]$,
$y_1 \ge -10$.

Here it is natural to look upon the index set as consisting of two different subsets since the constraints are generated by the vector

$$a(s) = (s,s^2)^T, \quad s \in [0,1],$$

$$a(2) = (1,0)^T.$$

(The notation a(2) is chosen arbitrarily.) The reader should verify that the constraints of (P) may be written in the form

$$a(s)^T y \ge b(s), \quad s \in S$$

where

$$S = [0,1] \cup \{2\}$$

and

$$b(s) = \begin{cases} 0, & s \in [0,1] \\ -10, & s = 2. \end{cases}$$

In the formulation of the corresponding dual problem we encounter infinitely many column vectors $a(s) \in R^2$. We may represent them in the "matrix" (see also §3)

$$\begin{pmatrix} 0 & \cdots & s & \cdots & 1 & 1 \\ 0 & \cdots & s^2 & \cdots & 1 & 0 \end{pmatrix}$$
$$\begin{array}{cccc} \uparrow & \uparrow & \uparrow & \uparrow \\ a(s) & a(s) & a(1) & a(2) \end{array}$$

$$s \in [0,1].$$

The dual problem can now be formulated at once. The constraints of (D) imply that the vector $(1,0)$ can be represented as a *nonnegative linear combination* of the vectors $a(s)$, $s \in S$:

$$\sum_{i=1}^{q-1} \begin{pmatrix} s_i \\ s_i^2 \end{pmatrix} x_i + \begin{pmatrix} 1 \\ 0 \end{pmatrix} x_q = \begin{pmatrix} 1 \\ 0 \end{pmatrix}, \quad x_1,\ldots,x_q \geq 0 \qquad (9)$$

$$s_1,\ldots,s_{q-1} \in [0,1]. \qquad (10)$$

The second of the two equations summarized in (9) is

$$\sum_{i=1}^{q-1} s_i^2 x_i = 0.$$

Because of (10) we must therefore have $x_i = 0$ or $s_i = 0$, $i = 1,\ldots,q-1$. Therefore $x_q = 1$ is necessary in order to satisfy (9) - (10). But then the value of the dual preference function becomes

$$\sum_{i=1}^{q} b(s_i)x_i = -10.$$

Thus we conclude

$$v(D) = -10.$$

We now determine $v(P)$. Since

$$sy_1 + s^2 y_2 \geq 0, \quad s \in [0,1],$$

we get $y_1 \geq 0$. ($sy_1 + s^2 y_2 = s(y_1 + sy_2)$ and $y_1 + sy_2 \geq 0$, *all* $s \in [0,1]$ implies $y_1 \geq 0$.)
    Therefore

$$0 \leq v(P).$$

We now note that every vector $(0, y_2)^T \in R^2$ with $y_2 \geq 0$ is optimal for
(P). Thus we conclude

$$v(P) = 0.$$

We have thus shown that the dual pair (P) - (D) has the duality gap

$$\delta(P,D) = 10.$$

Here we have an instance of Case 5 of the state diagram (1) or the defect
diagram (7) with $d = 10$. From this example we also realize that the defect d may be made arbitrarily large by appropriately choosing the constraints for (P).

(11) **Exercise.** Consider problem (7) of §4:

Minimize $y_1 + \frac{1}{2} y_2$ subject to $y_1 + sy_2 \geq e^s$, $s \in [0,1]$.

Show that both the primal problem and its dual are solvable and that no
duality gap occurs. **Hint:** Use for the dual $q = 2$ and $s_1 = 0$, $s_2 = 1$.

(12) Up to now we have not studied the *solvability* of (P) and (D).
This matter will be discussed in Chapter IV in connection with duality
theory.

(13) **Exercise.** a) Consider the linear optimization problem

(P)
$$\text{Minimize } -y_1 \text{ subject to the constraints } -y_1 \geq -1$$
$$-sy_1 - y_2 \geq 0, \quad s = 1,2,3,\ldots$$

Formulate the corresponding dual problem (D) and show that there is a
duality gap $\delta(P,D) = 1$.

b) Show that the problem (P) in a) is equivalent to the task:

Minimize $-y_1$ subject to $-y_1 \geq 0$

$$-y_1 - y_2 \geq 0.$$

Form the dual and show that no duality gap occurs.

(14) **Remark.** The example of the preceding exercise shows clearly
that the dual (D) of a certain linear optimization problem (P) depends
not only on the preference function and the *set* of feasible vectors but
also on the *formulation* of (P), i.e. on the manner in which the set of
feasible vectors is described through linear inequalities.

(15) **Exercise.** Consider again the Examples (2) and (5). The inequality $y_1 \geq 0$ is added to the constraints of (P) in (2). Show that

the corresponding dual pair is an instance of Case 5 of (1) and that no
duality gap occurs. Analogously, the inequality $0 \cdot y_1 \geq 1$ is added to
the constraints of Example (5). Show that the duality gap now "disappears"
(Case 6).

The question now arises whether the duality gap, when it occurs, is
caused by an "unfavorable" choice of inequalities

$$\sum_{r=1}^{n} a_r(s)y_r \geq b(s), \quad s \in S,$$

to describe the set of feasible vectors of (P). Is it possible that there
always is an equivalent system of inequalities

$$\sum_{r=1}^{n} \tilde{a}_r(s)y_r \geq \tilde{b}(s), \quad s \in S$$

describing the same set of vectors and such that *no* duality gap appears?

The answer is yes. The existence of an equivalent, but for the purpose of duality theory "better", system of inequalities is demonstrated
in a paper by Charnes, Cooper and Kortanek (1962). (See also Eckhardt
(1975).) However, there are no simple methods to transform systems of inequalities to remove duality gaps. Therefore we will not discuss these
questions further. Instead, we shall in Chapter IV give simple conditions
which insure that for a given linear optimization problem no duality gap
occurs.

Chapter III

Applications of Weak Duality in Uniform Approximation

Uniform approximation of functions is one of the most important applications of linear optimization. Both the theory and the computational treatment of linear optimization problems have been greatly influenced by the development of the theory of approximation.

In the first section of this chapter the general problem of uniform approximation will be formulated as a linear optimization problem. The corresponding dual is derived. The rest of the chapter will be devoted to the special case of polynomial approximation. Some classical problems which admit an exact solution in closed form are also studied.

§6. UNIFORM APPROXIMATION

Let T be an arbitrary set and $f: T \to R$ a real-valued function which is defined on T and bounded there. The real-valued bounded functions $v_r: T \to R$, $r = 1, \ldots, n$ are also given.

The problem of linear uniform approximation is to determine a linear combination

$$\sum_{r=1}^{n} y_r v_r$$

which best approximates f in the sense that the following expression is minimized:

$$\sup_{t \in T} \left| \sum_{r=1}^{n} y_r v_r(t) - f(t) \right|.$$

(1) The problem of uniform approximation:

(PA)

$$\text{Minimize} \quad \sup_{t \in T} \left| \sum_{r=1}^{n} y_r v_r(t) - f(t) \right|$$

over all vectors $y = (y_1, \ldots, y_n)^T \in R^n$.

An equivalent formulation is

$$\text{Minimize} \quad y_{n+1} \quad \text{over all vectors} \quad (y, y_{n+1})^T \in R^{n+1},$$

subject to the constraints

$$\left| \sum_{r=1}^{n} y_r v_r(t) - f(t) \right| \le y_{n+1}, \quad \textit{all} \quad t \in T.$$

We note that for real numbers α and β the inequality

$$|\alpha| \le \beta$$

is equivalent to the two inequalities

$$-\alpha \ge -\beta$$
$$\alpha \ge -\beta$$

Therefore the approximation problem (PA) may be rewritten in the following form:

Minimize y_{n+1} subject to the constraints   (2)

$$\sum_{r=1}^{n} v_r(t) y_r + y_{n+1} \ge f(t), \quad \text{all} \quad t \in T \qquad (3)$$

$$-\sum_{r=1}^{n} v_r(t) y_r + y_{n+1} \ge -f(t), \quad \text{all} \quad t \in T. \qquad (4)$$

This problem now has the form of a linear optimization problem (P) in R^{n+1} provided the index set S and the functions $a(s) = (a_1(s), \ldots, a_n(s))^T$ are properly defined. There are two different kinds of vectors $a(s)$ since the vectors

$$\begin{pmatrix} v_1(t) \\ \vdots \\ v_n(t) \\ 1 \end{pmatrix} \quad \text{and} \quad \begin{pmatrix} -v_1(t) \\ \vdots \\ -v_n(t) \\ 1 \end{pmatrix}, \quad t \in T, \qquad (5)$$

correspond to the conditions (3) and (4) respectively. The constraints of the *dual* of the problem (2) - (4) imply that the vector

$$c = \begin{pmatrix} 0 \\ \vdots \\ 0 \\ 1 \end{pmatrix} \in R^{n+1},$$

which appears in the preference function of (2), must be expressed as a nonnegative linear combination of finitely many of the vectors (5).

Hence the *dual problem* corresponding to (2) - (4) takes the form (compare with §4, (15) - (17)):

Determine two subsets $\{t_1^+,\ldots,t_{q^+}^+\}$, $\{t_1^-,\ldots,t_{q^-}^-\}$ of $T(q^+ + q^- \geq 1)$ and real numbers $x_1^+,\ldots,x_{q^+}^+$, $x_1^-,\ldots,x_{q^-}^-$ such that the expression

$$\sum_{i=1}^{q^+} f(t_i^+)x_i^+ - \sum_{i=1}^{q^-} f(t_i^-)x_i^- \tag{6}$$

is maximized, subject to the constraints

$$\sum_{i=1}^{q^+} v_r(t_i^+)x_i^+ - \sum_{i=1}^{q^-} v_r(t_i^-)x_i^- = 0, \quad r = 1,\ldots,n, \tag{7}$$

$$\sum_{i=1}^{q^+} x_i^+ + \sum_{i=1}^{q^-} x_i^- = 1, \tag{8}$$

$$x_i^+ \geq 0, \quad i = 1,\ldots,q^+, \tag{9}$$

$$x_i^- \geq 0, \quad i = 1,\ldots,q^-. \tag{10}$$

This dual problem can be written in an equivalent, but simpler form.

(11) **The dual problem (DA)**. Determine a subset $\{t_1,\ldots,t_q\}$ of T $(q \geq 1)$ and real numbers $x_1,x_2,\ldots,x_q$ such that the expression

$$\sum_{i=1}^{q} f(t_i)x_i \tag{12}$$

is maximized, subject to the constraints

$$\sum_{i=1}^{q} v_r(t_i)x_i = 0, \quad r = 1,\ldots,n, \tag{13}$$

$$\sum_{i=1}^{q} |x_i| \leq 1. \tag{14}$$

(15) **Lemma**. The optimization problems (6) - (10) and (12) - (14) are equivalent in the following sense: For every $\{t_1^+,\ldots,t_{q^+}^+, t_1^-,\ldots,t_{q^-}^-, x_1^+,\ldots,x_{q^+}^+, x_1^-,\ldots,x_{q^-}^-\}$ satisfying (7) - (10) one may construct

$\{t_1,\ldots,t_q,\ x_1,\ldots,x_q\}$ satisfying (13), (14) such that the values of the preference functions (6) and (12) coincide, and vice-versa.

Proof: Let a solution of (7) - (10) be given. We may as well assume that $x_i^+ > 0$, $x_i^- > 0$. We discuss first the case when the sets $T^+ = \{t_1^+,\ldots,t_{q^+}^+\}$ and $T^- = \{t_1^-,\ldots,t_{q^-}^-\}$ are *disjoint*. Then we just put $q = q^+ + q^-$, $\{t_1,\ldots,t_q\} = T^+ \cup T^-$ and

$$x_i = \begin{cases} x_j^+, & \text{if } t_i = t_j^+ \text{ for a } t_j^+ \in T^+ \\ -x_j^-, & \text{if } t_i = t_j^- \text{ for a } t_j^- \in T^-, \quad i = 1,\ldots,q. \end{cases}$$

It is easy to verify that (13), (14) are satisfied and that (6) and (12) have the same value. In the remaining case when T^+ and T^- have a point in common, there are indices k, ℓ such that

$$t_k^+ = t_\ell^-, \quad \min(x_k^+, x_\ell^-) = d > 0.$$

Then we replace x_k^+ with $x_k^+ - d$ and x_i^- with $x_i^- - d$. If now $x_k^+ - d = 0$ then we remove t_k^+ from T^+, but if instead $x_i^- - d = 0$, t_k^- is removed from T^-. This transformation does not change the value of the preference function (6), and the equations (7), (9), (10) continue to hold. But instead of (8) we get

$$\sum_{i=1}^{q^+} x_i^+ + \sum_{i=1}^{q^-} x_i^- \leq 1.$$

The sets T^+ and T^- will become disjoint after a finite number of the transformations described above and a suitable solution of (DA) is constructed by the procedure given earlier. To verify the remaining part of the assertion we let $\{t_1,\ldots,t_q,\ x_1,\ldots,x_q\}$ be feasible for (DA). Now set $q^+ = q^- = q$, $t_i^+ = t_i$, $i = 1,\ldots,q$, and

$$x_i^+ = \max(0,x_i) = (|x_i| + x_i)/2,$$

$$x_i^- = -\min(0,x_i) = (|x_i| - x_i)/2, \quad i = 1,\ldots,q.$$

The rest of the argument is straightforward. Note that in order to satisfy (8) it might be necessary to replace x_i^+ with $x_i^+ + c$, x_i^- with $x_i^- + c$, where $c \geq 0$ is chosen so that the condition (8) is met.

All duality results which have been derived for the dual pair (2) - (4), (6) - (10) may be applied to the pair of problems (PA), (DA) from (1) and (11) to give corresponding statements. However, many of these

theorems may be shown directly for the pair (PA) - (DA). This is true,
e.g. for the duality lemma which could be based on (1) of §4:

(16) **Lemma.** Let the finite subset $\{t_1,\ldots,t_q\} \subset T$ and the real
numbers $x_1,\ldots,x_q$ be such that

$$\sum_{i=1}^{q} v_r(t_i)x_i = 0, \quad r = 1,\ldots,n \tag{17}$$

$$\sum_{i=1}^{q} |x_i| \leq 1. \tag{18}$$

Then the following relation holds for any $y \in R^n$:

$$\sum_{i=1}^{q} f(t_i)x_i \leq \sup_{t \in T} |\sum_{r=1}^{n} y_r v_r(t) - f(t)|. \tag{19}$$

Proof: From (17) we conclude

$$\sum_{i=1}^{q} \left(\sum_{r=1}^{n} y_r v_r(t_i)\right) x_i = 0.$$

Thus

$$\sum_{i=1}^{q} f(t_i)x_i = \sum_{i=1}^{q} \left\{ f(t_i) - \sum_{r=1}^{n} y_r v_r(t_i) \right\} x_i$$

$$\leq \sum_{i=1}^{q} |f(t_i) - \sum_{r=1}^{n} y_r v_r(t_i)| \, |x_i|$$

$$\leq \sup_{t \in T} |f(t) - \sum_{r=1}^{n} y_r v_r(t)| \sum_{i=1}^{q} |x_i|$$

$$\leq \sup_{t \in T} |f(t) - \sum_{r=1}^{n} y_r v_r(t)|$$

which is the desired result.

(20) **Exercise.** Show that the left hand side of (19) may be replaced
by

$$|\sum_{i=1}^{q} f(t_i)x_i|.$$

(21) **Remark.** If $q \geq n+1$, then (17) has a nontrivial solution for
any choice of elements $t_1,\ldots,t_q$ in T. Indeed, (17) then gives the
underdetermined linear system of equations

$$
\begin{pmatrix}
v_1(t_1) & \cdots & v_1(t_q) \\
v_2(t_1) & \cdots & v_2(t_q) \\
\cdot & & \cdot \\
\cdot & & \cdot \\
\cdot & & \cdot \\
v_n(t_1) & & v_n(t_q)
\end{pmatrix}
\begin{pmatrix}
\hat{x}_1 \\
\hat{x}_2 \\
\cdot \\
\cdot \\
\cdot \\
\hat{x}_q
\end{pmatrix}
=
\begin{pmatrix}
0 \\
0 \\
\cdot \\
\cdot \\
\cdot \\
0
\end{pmatrix},
$$

and setting

$$
x = \left(\sum_{i=1}^{q} |\hat{x}_i| \right)^{-1} \hat{x}, \tag{22}
$$

the vector $x \in R^q$ now meets the constraints (17), (18) of (DA).

(23) <u>Example</u>. The function $f(t) = e^t$ is to be uniformly approximated by a straight line $y_1 + y_2 t$ over the interval $T = [-1,1]$. Thus we need to solve the problem:

Minimize $\sup_{t \in T} |e^t - y_1 - y_2 t|$.
(y_1, y_2)

We want to apply Lemma (16). We select $q = 3$ and set $t_1 = -1$, $t_2 = 0$, $t_3 = 1$. The system of equations (17) then becomes

$$
\hat{x}_1 + \hat{x}_2 + \hat{x}_3 = 0
$$
$$
-\hat{x}_1 \qquad + \hat{x}_3 = 0.
$$

The general solution of this system is given by

$$
\hat{x}_1 = \alpha
$$
$$
\hat{x}_2 = -2\alpha
$$
$$
\hat{x}_3 = \alpha
$$

where α is arbitrary. The "normalization" (22) gives

$$
x = (\tfrac{1}{4}, -\tfrac{1}{2}, \tfrac{1}{4})^T,
$$

which together with $t_1 = -1$, $t_2 = 0$, $t_3 = 1$ meets the constraints of (DA). Thus we may conclude from (16) that if e^t is approximated by a straight line over the interval $[-1,1]$, then the error will be *at least*

$$
\tfrac{1}{4}e^{-1} - \tfrac{1}{2} + \tfrac{1}{4}e \approx 0.27.
$$

An *upper bound* for the smallest possible approximation error is obtained by taking

$$y_1 + y_2 t = 1.36 + t.$$

Then

$$\sup_{t \in [-1,1]} |e^t - 1.36 - t| \approx 0.36.$$

(24) <u>Exercise</u>. The function $f(t) = 1/(2+t)$ is to be uniformly approximated over the interval $[-1,1]$ by a straight line $y_1 + y_2 t$. Determine a lower bound for the value of the corresponding approximation problem by proceeding as in (23). (Thus one selects again $q = 3$ and puts $t_1 = -1$, $t_2 = 0$, $t_3 = 1$.) <u>Hint</u>: One gets the same linear system for x_1, x_2, x_3 as in the preceding example. (The lower bound is 0.083.)

(25) <u>Exercise</u>. Consider the same example as in (24) with $q = 3$. Let $t_1 = -1$, $t_3 = 1$, but set $t_2 = \tau$. Then try to determine $t_2 = \tau$ *optimally* for (DA). <u>Hint</u>: x_1, x_2, x_3 and τ become the variables of the following optimization problem:

$$\text{Maximize } x_1 + \frac{x_2}{1+\tau} + \frac{x_3}{3}$$

subject to the constraints

$$x_1 + x_2 + x_3 = 0,$$

$$-x_1 + \tau x_2 + x_3 = 0, \tag{26}$$

$$|x_1| + |x_2| + |x_3| = 1, \tag{27}$$

$$-1 \le \tau \le 1. \tag{28}$$

Assume that x_1 and x_3 are positive and x_2 negative. Then (27) becomes $x_1 - x_2 + x_3 = 1$. This relation is used together with (26) to express x_1, x_2 and x_3 as (linear) functions of τ. We then enter these expressions into the preference function and maximize with respect to τ. This gives the lower bound 0.0893.

The following simple lemma may be useful when one wants to show that a certain vector y is an optimal solution of (PA). An illustrative example is given in (31).

(29) <u>Lemma</u>. Let $\{t_1, \ldots, t_q, x_1, \ldots, x_q\}$, where $t_i \in T$, $i = 1, \ldots, q$, and $q \ge 1$, satisfy

$$\sum_{i=1}^{q} v_r(t_i) x_i = 0, \quad r = 1, \ldots, n,$$

$$\sum_{i=1}^{q} |x_i| = 1.$$

Let $y \in R^n$ and define

$$y_{n+1} = \sup_{t \in T} \left| f(t) - \sum_{r=1}^{n} y_r v_r(t) \right|.$$

Assume also that the following relations hold for $i = 1,\ldots,q$: Either $x_i = 0$ or

$$f(t_i) - \sum_{r=1}^{n} y_r v_r(t_i) = y_{n+1} \, \text{sgn} \, x_i \quad \text{where} \quad \text{sgn} \, x_i = x_i / |x_i|. \qquad (30)$$

Then we may assert: $\{t_1,\ldots,t_q, x_1,\ldots,x_q\}$ is an optimal solution of (DA) and y of (PA), and the values of (PA) and (DA) coincide.

<u>Proof:</u>

$$\sum_{i=1}^{q} f(t_i) x_i = \sum_{i=1}^{q} f(t_i) x_i - \sum_{r=1}^{n} y_r \left(\sum_{i=1}^{q} v_r(t_i) x_i \right)$$

$$= \sum_{i=1}^{q} \left\{ f(t_i) - \sum_{r=1}^{n} y_r v_r(t_i) \right\} x_i.$$

Applying (30) we get

$$\sum_{i=1}^{q} f(t_i) x_i = y_{n+1} \sum_{i=1}^{q} x_i \, \text{sgn}(x_i) = y_{n+1} \sum_{i=1}^{q} |x_i|$$

$$= \sup_{t \in T} \left| f(t) - \sum_{r=1}^{n} y_r v_r(t) \right|.$$

The statement now follows from Lemma (16).

(31) <u>Example.</u> The function $f(t) = t^2$ is to be uniformly approximated over the interval $[0,2]$ with a linear combination of the functions $v_1(t) = t$, $v_2(t) = \exp(t)$.

Andreasson and Watson (1976) give as the solution of this approximation problem the following coefficients y_1 and y_2 of v_1 and v_2:

$$y_1 = 0.18423256, \quad y_2 = 0.41863122.$$

We want to use Lemma (29) to verify that these values of y_1 and y_2 are optimal (within the precision shown). One first establishes that the error function

$$t^2 - y_1 t - y_2 \exp(t)$$

assumes its minimum and maximum values at $t_1 = 0.40637574$ and $t_2 = 2.00000000$:

$$t_1^2 - y_1 t_1 - y_2 \exp(t_1) = -0.53824531,$$
$$t_2^2 - y_1 t_2 - y_2 \exp(t_2) = 0.53824531.$$

The dual constraints from (29) read (with $q = 2$)

$$t_1 x_1 + t_2 x_2 = 0,$$
$$\exp(t_1) x_1 + \exp(t_2) x_2 = 0,$$
$$|x_1| + |x_2| = 1.$$

We put sgn $x_1 = 1$ and sgn $x_2 = -1$. Then two of the equations above become

$$t_1 x_1 + t_2 x_2 = 0,$$
$$-x_1 + x_2 = 1.$$

Entering $t_1 = 0.40637574$ and $t_2 = 2$ into these equations we obtain $x_1 = -0.83112540$ and $x_2 = 0.16887459$. It is now easy to check that all conditions of Lemma (29) are met. Thus the proposed solution is indeed optimal.

We conclude this section by showing that the approximation problem is solvable under fairly general conditions.

(32) **Theorem.** Let $T \subset R^k$ be nonempty and compact and assume also that the functions f, $v_1, \ldots, v_n$ are continuous and linearly independent on T. Then the linear approximation problem (PA) is solvable; i.e. there is a vector $\hat{y} \in R^n$ such that

$$\max_{t \in T} \left| f(t) - \sum_{r=1}^{n} \hat{y}_r v_r(t) \right| = \min_{y \in R^n} \max_{t \in T} \left| f(t) - \sum_{r=1}^{n} y_r v_r(t) \right|.$$

Note. We may write "max" instead of "sup" in the formulation of (PA) since the functions f, $v_1, \ldots, v_n$ are continuous and T is compact and hence the error function

$$f - \sum_{r=1}^{n} y_r v_r$$

assumes its maximum and its minimum.

Proof: We define a norm on R^n by

$$||y||_v = \max_{t \in T} \left| \sum_{r=1}^{n} y_r v_r(t) \right|.$$

Putting $y = 0$ we get

$$\max_{t \in T} \left| f(t) - \sum_{r=1}^{n} y_r v_r(t) \right| = \max_{t \in T} \left| f(t) \right| = \Delta.$$

Hence the optimum value of (PA) lies in the interval $[0, \Delta]$. Because of the minimization we need only to consider those vectors y which satisfy

$$\max_{t \in T} \left| f(t) - \sum_{r=1}^{n} y_r v_r(t) \right| \leq \Delta. \tag{33}$$

Using the triangle inequality on (33) we find

$$\left| \sum_{r=1}^{n} y_r v_r(t) \right| \leq \left| f(t) - \sum_{r=1}^{n} y_r v_r(t) \right| + \left| f(t) \right| \leq 2\Delta.$$

Thus we need only to minimize over those vectors $y \in R^n$ such that

$$\|y\|_v \leq 2\Delta;$$

i.e. a compact subset of R^n. Since the preference function of (PA),

$$y \to \max_{t \in T} \left| f(t) - \sum_{r=1}^{n} y_r v_r(t) \right|,$$

is continuous, the existence of an optimal solution follows by Weierstrass' theorem (see (13), §2).

§7. POLYNOMIAL APPROXIMATION

This section is devoted to the study of (PA) in the case when T is a real interval and the function f is to be approximated by a polynomial. Then major simplifications are possible and one can, for example, calculate lower bounds for the error of the best approximation without treating the dual problem explicitly. Some special approximation problems admitting an optimal solution in closed form are also treated. We now prove:

(1) **Lemma.** Let $t_1 < t_2 < \ldots < t_{n+1}$ be fixed real numbers and let $(x_1, \ldots, x_{n+1})$ be a nontrivial solution of the homogeneous linear system of equations

$$\sum_{i=1}^{n+1} t_i^{r-1} x_i = 0, \quad r = 1, \ldots, n. \tag{2}$$

Then

$$x_i x_{i+1} < 0, \quad i = 1, \ldots, n.$$

Proof: Let i be a fixed integer such that $1 \leq i \leq n$. Denote by P_n the uniquely determined polynomial

$$P_n(t) = \sum_{r=1}^{n} y_r t^{r-1}$$

satisfying

$$P_n(t_j) = \begin{cases} 1, & j = i \\ 0, & j = 1,\ldots,n+1, \quad j \neq i, \quad j \neq i+1. \end{cases}$$

(See Fig. 7.1.) That such a P_n does exist is an immediate consequence of the fact that the so-called *Vandermonde* matrix is nonsingular. (See (3) below.) From (2),

$$\sum_{i=1}^{n+1} P_n(t_i) x_i = \sum_{r=1}^{n} y_r \sum_{i=1}^{n+1} t_i^{r-1} x_i = 0.$$

Due to the construction of P_n this relation gives

$$x_i + P_n(t_{i+1}) x_{i+1} = 0.$$

P_n cannot vanish in $[t_i, t_{i+1}]$; if it did, P_n would have n zeros, which is impossible. Therefore $P_n(t_{i+1}) > 0$ and we conclude $x_i x_{i+1} < 0$.

(3) **Exercise.** Let $t_1 < t_2 < \ldots < t_n$ be given. Define the *Vandermonde* matrix V by

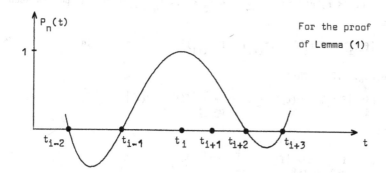

For the proof
of Lemma (1)

Fig. 7.1

$$V(t_1,\ldots,t_n) = \begin{pmatrix} 1 & 1 & \cdots & 1 \\ t_1 & t_2 & \cdots & t_n \\ t_1^2 & t_2^2 & \cdots & t_n^2 \\ \vdots & \vdots & & \vdots \\ t_1^{n-1} & t_2^{n-1} & \cdots & t_n^{n-1} \end{pmatrix}.$$

It can be shown that

$$\det V(t_1,\ldots,t_n) > 0. \tag{4}$$

Use (2) to obtain the expression

$$x_i = -x_{i+1} \frac{\det V(t_1,\ldots,t_{i-1},\ t_{i+1},\ldots,t_{n+1})}{\det V(t_1,\ldots,t_{i-1},\ t_i,\ t_{i+2},\ldots,t_{n+1})}.$$

This combined with (4) gives an alternative proof of Lemma (1).

We remark here that a result corresponding to Lemma (1) may be established not only for $1,t,\ldots,t^{n-1}$, but also for general *Chebyshev systems* $v_1,\ldots,v_n$. The theorems to follow which depend on Lemma (1) can also be generalized. See Chapter VIII.

The following theorem, which is due to De La Vallee-Poussin, is important since it can be used for calculating lower bounds for the error of the best possible approximation without solving the linear system (2) explicitly.

(5) **Theorem.** Let f be continuous on $[\alpha,\beta]$, let P be a polynomial of degree $< n$, and let $\alpha \le t_1 < t_2 < \ldots < t_{n+1} \le \beta$ be n points such that

$$\{f(t_i) - P(t_i)\} \cdot \{f(t_{i+1}) - P(t_{i+1})\} < 0, \quad i = 1,\ldots,n. \tag{6}$$

(See Fig. 7.2.) Then

$$\min_i |f(t_i) - P(t_i)| \le \Delta_n \le \max_{\alpha \le t \le \beta} |f(t) - P(t)|, \tag{7}$$

where

$$\Delta_n = \inf_{y \in R^n} \max_{\alpha \le t \le \beta} |f(t) - \sum_{r=1}^{n} y_r t^{r-1}|.$$

Thus Δ_n denotes the smallest error which can be achieved when f is approximated by polynomials of degree $< n$.

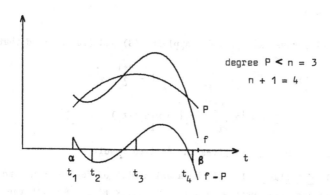

Fig. 7.2

Proof: The right-hand inequality in (7) is obvious. Let $\rho_1, \ldots, \rho_{n+1}$ be a nontrivial solution of the system

$$\sum_{i=1}^{n+1} t_i^{r-1} \rho_i = 0, \quad r = 1, \ldots, n.$$

By Lemma (1) we may assume $\rho_i \rho_{i+1} < 0$, $i = 1, \ldots, n$. Now put

$$x_i = \rho_i \left\{ \sum_{j=1}^{n+1} |\rho_j| \right\}^{-1}.$$

In this way we get a feasible solution to the dual problem since

$$\sum_{i=1}^{n+1} t_i^{r-1} x_i = 0, \quad r = 1, \ldots, n,$$

$$\sum_{i=1}^{n+1} |x_i| = 1. \tag{8}$$

By (16) of §6 (weak duality) we also have

$$\sum_{i=1}^{n+1} f(t_i) x_i \leq \Delta_n. \tag{9}$$

We now define

$$\delta_i = f(t_i) - P(t_i);$$

by assumption (6), $\delta_i \delta_{i+1} < 0$. If the signs of all numbers x_i are changed simultaneously, the constraints of (DA) are still met. Therefore we can always achieve

$$x_i \delta_i > 0 \tag{10}$$

since we also have $x_i x_{i+1} < 0$. Applying (8) and (10) we find that

$$\sum_{i=1}^{n+1} f(t_i) x_i = \sum_{i=1}^{n+1} x_i |f(t_i) - P(t_i)| = \sum_{i=1}^{n+1} x_i \delta_i$$

$$\geq \min_i |\delta_i| \sum_{i=1}^{n+1} |x_i| = \min_i |f(t_i) - P(t_i)|.$$

An application of (9) now gives the desired result.

(11) <u>Corollary</u>. Let P be a polynomial of degree $< n$ and such that there are $n+1$ points $\alpha \leq t_1 < t_2 < \ldots < t_{n+1} \leq \beta$ with the properties

$$|\delta_i| = |f(t_i) - P(t_i)| = \max_{\alpha \leq t \leq \beta} |f(t) - P(t)|, \quad i = 1, \ldots, n+1, \tag{12}$$

$$\text{and} \quad \delta_i \delta_{i+1} < 0, \quad i = 1, \ldots, n.$$

Then P is a polynomial of degree $< n$ which best approximates f in the uniform norm. The conditions (12) state that the error function $f - P$ alternates in sign at $t_1, \ldots, t_{n+1}$ and assumes its largest absolute value at these points.

(13) <u>Remark</u>. In the special case when

$$|\delta_1| = |\delta_2| = \ldots = |\delta_{n+1}|,$$

we get

$$\sum_{i=1}^{n+1} f(t_i) x_i = \min_i |f(t_i) - P(t_i)|.$$

Hence (7) and (9) give the same lower bound for the attainable approximation error in this case. We shall show in Chapter IV that a strong duality theorem can be established for the dual pair (PA) and (DA); i.e. no duality gap occurs. This entails the use of Theorem (5) for constructing *arbitrarily good* lower bounds for Δ_n by choosing $t_1, t_2, \ldots, t_{n+1}$ suitably.

(14) <u>Determination of a polynomial satisfying (6)</u>. Let $\alpha \leq t_1 < t_2 < \ldots < t_{n+1} \leq \beta$ be given. Define the function δ by

$$\delta(t_i) = (-1)^i, \quad i = 1, \ldots, n+1.$$

We now seek a polynomial P of degree $< n$ and a constant ε such that

$$P(t_i) = f(t_i) + \varepsilon\delta(t_i), \quad i = 1,\ldots,n+1. \tag{15}$$

(15) is a linear system of equations with ε and the coefficients of P as unknowns. Using (4) it is easy to demonstrate that P and ε are uniquely determined by (15).

P and ε are efficiently calculated using a so-called difference scheme. (We assume that divided differences are familiar to the reader. Otherwise see e.g. Dahlquist and Björck, (1974), p. 277.) Since

$$P[t_1,\ldots,t_{n+1}] = 0,$$

(15) gives at once

$$\varepsilon = -f[t_1,\ldots,t_{n+1}]/\delta[t_1,\ldots,t_{n+1}],$$

where we use the customary notations for divided differences. P may be represented in the "Newton" form

$$P(t) = P[t_1] + P[t_1,t_2](t-t_1) + \ldots + P[t_1,t_2,\ldots,t_n] \prod_{i=1}^{n-1} (t-t_i).$$

The divided differences appearing in this formula are easily computed from the intermediate results obtained when calculating ε. By Theorem (5), $|\varepsilon|$ is a lower bound for Δ_n.

(16) <u>Numerical example</u>. Let $[\alpha,\beta] = [0,1]$, $f(t) = (1+t)^{-1}$, $n = 2$, $t_1 = 0$, $t_2 = 1/2$, and $t_3 = 1$. The difference schemes for f and δ are:

t_i	$f(t_i)$	$f[t_i,t_{i+1}]$	$f[t_1,t_2,t_3]$	$\delta(t_i)$	$\delta[t_i,t_{i+1}]$	$\delta[t_1,t_2,t_3]$
0	1			-1		
		-2/3			4	
1/2	2/3		1/3	1		-8
		-1/3			-4	
1	1/2			-1		

We get at once $\varepsilon = 1/24$; i.e. the function $1/(1+t)$ cannot be approximated in the uniform norm over $[0,1]$ by a straight line with an error less than $1/24 \approx 0.0417$.

(17) <u>Exercise</u>. Take $t_1 = 0$, $t_3 = 1$, and show by optimizing over $t_2 \in (0,1)$ (see also exercise (25), §6) that $\Delta_2 = (3-\sqrt{8})/4 \approx 0.0429$.

We now discuss some special approximation problems which nevertheless are of general interest.

(18) <u>Exercise</u>. Let f have two continuous derivatives on $[\alpha,\beta]$ and be such that $f''(t) > 0$, $t \in [\alpha,\beta]$. Denote by ℓ the straight line which interpolates f at the endpoints α and β. Put

$$\delta = \max_{\alpha \leq t \leq \beta} |f(t) - \ell(t)|.$$

Next use (11) to show that the straight line which approximates f best in the uniform norm has the representation

$$\ell(t) - \delta/2$$

and that the approximation error is $\delta/2$.

(19) <u>Exercise</u>. Put $f(t) = t^2$ in (18) and show that the straight line which best approximates $f(t) = t^2$ in the uniform norm interpolates this function at $\hat{t}_1$ and $\hat{t}_2$, where

$$\hat{t}_1 = \frac{\alpha+\beta}{2} - \frac{1}{2\sqrt{2}} (\beta-\alpha), \qquad \hat{t}_2 = \frac{\alpha+\beta}{2} + \frac{1}{2\sqrt{2}} (\beta-\alpha).$$

Show also that the approximation error is

$$\frac{1}{8} (\alpha-\beta)^2.$$

We will next treat the more general problem of approximating $f(t) = t^n$ in the uniform norm by a polynomial of degree < n. In order to represent the solution in a concise form we introduce the Chebyshev polynomials.

(20) <u>Definition</u>. The *Chebyshev polynomials* $T_0, T_1, \ldots$ are defined through

$$\begin{aligned} &T_0(t) = 1, \quad T_1(t) = t \\ &T_n(t) = 2t\, T_{n-1}(t) - T_{n-2}(t), \quad n = 2,3,\ldots \end{aligned} \tag{21}$$

(22) <u>Exercise</u>. Show that the recurrence relation (21) is satisfied by

$$T_n(t) = \cos(n \arccos t). \tag{23}$$

<u>Hint</u>: Use the addition theorem

$$\cos(A+B) = 2 \cos A \cos B - \cos (A-B). \tag{24}$$

We now prove:

(25) <u>Theorem</u>. Let $[\alpha,\beta]$ be a given interval. Then

$$\min_{t_1,\ldots,t_n} \; \max_{\alpha \leq t \leq \beta} \; \left| \prod_{i=1}^{n} (t-t_i) \right| = 2(\beta-\alpha)^n/4^n.$$

The minimum is assumed for

$$\hat{t}_i = \frac{\alpha+\beta}{2} + \frac{\beta-\alpha}{2} \cos \theta_i, \quad \text{where} \quad \theta_i = \frac{i-1/2}{n} \pi, \quad i = 1,\ldots,n. \qquad (26)$$

Also,

$$\prod_{i=1}^{n} (t-\hat{t}_i) = \frac{2(\beta-\alpha)^n}{4^n} T_n(\frac{2t-\alpha-\beta}{\beta-\alpha}).$$

Proof: Consider the approximation problem

$$\underset{\substack{y \in R^n}}{\text{Minimize}} \ \max_{t \in [\alpha,\beta]} \left| t^n - \sum_{r=1}^{n} y_r t^{r-1} \right|. \qquad (27)$$

We next determine y through the condition

$$t^n - \sum_{r=1}^{n} y_r t^{r-1} = Q_n(t), \qquad (28)$$

where

$$Q_n(t) = \frac{2(\beta-\alpha)^n}{4^n} T_n(\frac{2t-\alpha-\beta}{\beta-\alpha}), \qquad (29)$$

and apply (11) to verify that y is a solution of (27).

We first note that

$$t \to \frac{2t-\alpha-\beta}{\beta-\alpha}$$

maps $[\alpha,\beta]$ on $[-1,1]$. Using the recurrence relation (21) for T_n, we verify that the coefficient of t^n in Q_n is 1. By (23) we conclude that $|T_n(t)| \leq 1$ and thus $|Q_n(t)| \leq 2(\beta-\alpha)^n/4^n$. We also find that

$$Q_n(t_i^*) = (-1)^{i-1} 2(\beta-\alpha)^n/4^n,$$

where

$$t_i^* = \frac{\alpha+\beta}{2} + \frac{\beta-\alpha}{2} \cos \frac{(i-1)\pi}{n}, \quad i = 1,\ldots,n+1, \qquad (30)$$

and $|Q_n(t)|$ assumes its maximum value at t_i^*. Hence the conclusion follows from (11).

Using (28) and (29) we conclude that the polynomial P_n of degree $< n$ which best approximates t^n in the uniform norm on $[\alpha,\beta]$ is given by

$$P_n(t) = t^n - \frac{2(\beta-\alpha)^n}{4^n} T_n(\frac{2t-\alpha-\beta}{\beta-\alpha}). \qquad (31)$$

(Note that when the right hand side is expanded the coefficient of t^n vanishes.)

(32) Exercise. Let

$$t_i = \cos \frac{i-1}{N} \pi, \quad i = 1,\ldots,N+1,$$

and

$$s_i = \cos \frac{i-1/2}{N}, \quad i = 1,\ldots,N.$$

Show that the Chebyshev polynomials satisfy the following orthogonality relations:

$$\frac{1}{N} \sum_{i=1}^{N} T_m(s_i)T_n(s_i) = \begin{cases} 1, & m = n = 0 \\ 1/2, & 0 < m = n < N \\ 0, & m \neq n, \ m < N, \ n \leq N, \end{cases} \tag{33}$$

$$\frac{1}{N} \sum_{i=1}^{N+1} {}''T_m(t_i)T_n(t_i) = \begin{cases} 1, & m = n = 0 \\ 1/2, & 0 < m = n < N \\ 0, & m \neq n, \ m \leq N, \ n \leq N. \end{cases} \tag{34}$$

Here the notation " means that the factor $1/2$ should be placed in front of the first and the last term in (34). Note also that $T_N(t_i) = (-1)^{i-1}$ and $T_N(s_i) = 0$.

We next treat an approximation problem which sometimes occurs in the study of iteration processes in numerical linear algebra.

(35) Theorem. Let $[\alpha,\beta]$ be a bounded interval such that $0 \notin [\alpha,\beta]$. Consider the problem

$$\text{Minimize} \quad \max_{\alpha \leq t \leq \beta} |P(t)| \tag{36}$$

over all polynomials of degree $\leq n$ such that

$$P(0) = 1. \tag{37}$$

The optimal solution is given by

$$P(t) = T_n\left(\frac{2t-\alpha-\beta}{\beta-\alpha}\right)/T_n\left(\frac{\alpha+\beta}{\alpha-\beta}\right). \tag{38}$$

Proof: We can write P in the form

$$P(t) = 1 - y_1 t - y_2 t^2 - \ldots - y_n t^n \tag{39}$$

since $P(0) = 1$. The problem (36), (37) may then be written

$$\min_{y_1,\ldots,y_n} \quad \max_{\alpha \leq t \leq \beta} |1 - y_1 t - \ldots - y_n t^n| \tag{40}$$

and we recognize (40) as an instance of (PA) in (1) of §6. Its dual reads: determine subsets $\{t_1,\ldots,t_q\} \subset [\alpha,\beta]$ $(q \geq 1)$, and real numbers $x_1,\ldots,x_q$ such that

$$\sum_{i=1}^{q} x_i \tag{41}$$

is maximized subject to the constraints

$$\sum_{i=1}^{q} x_i t_i^{r-1} = 0, \quad r = 2,\ldots,n+1 \tag{42}$$

$$\sum_{i=1}^{q} |x_i| \leq 1. \tag{43}$$

See (11) of §6. We shall construct feasible solutions to the two problems (40) and (41) - (43), and then use (29) of §6 to verify that these solutions are optimal.

In (41) - (43) we put $q = n+1$,

$$t_i = \frac{\beta+\alpha}{2} + \frac{\beta-\alpha}{2} \cos \theta_i, \quad \theta_i = \frac{i-1}{n} \pi, \quad i = 1,\ldots,n+1$$

$$x_1 = \frac{1}{2n} T_n(\cos \theta_1), \quad x_{n+1} = \frac{1}{2n} T_n(\cos \theta_{n+1}) \tag{44}$$

$$x_i = \frac{1}{n} T_n(\cos \theta_i), \quad i = 2,\ldots,n.$$

Condition (42) is now met by (33) since we may express t_i^{r-1} as a linear combination of $T_0,\ldots,T_{r-1}$. Since $|T_n(t_i)| = 1$, (43) is also satisfied. We observe that P in (38) is of the form (39). Next we define $y_1,\ldots,y_n$ by (39) for this particular polynomial P. By (38),

$$P(t_i) = (-1)^{i-1}/T_n(\frac{\alpha+\beta}{\alpha-\beta}).$$

Now (44) gives

$$x_1 = \frac{1}{2n}, \quad x_{n+1} = \frac{(-1)^n}{2n}, \quad x_i = \frac{(-1)^{i-1}}{n}, \quad i = 2,\ldots,n.$$

Hence (30) of §6 is also met, establishing optimality of the polynomial (38).

We next discuss the problem of constructing polynomials of degree $< n$ which approximate a function f on a bounded interval $[\alpha,\beta]$. Two approaches are conceivable: i) select n points $t_1 < t_2 <\ldots< t_n$ and determine the polynomial P of degree $< n$ satisfying

$$P(t_i) = f(t_i), \quad i = 1,2,\ldots,n; \tag{45}$$

ii) select n+1 points $s_1 < s_2 < ... < s_{n+1}$ and determine the polynomial
Q of degree < n which solves the problem

$$\min_{Q} \max_{s_1,...,s_{n+1}} |f(t) - Q(t)|, \quad \text{degree } Q < n. \tag{46}$$

(47) <u>Exercise</u>. Show that there is one and only one polynomial P
of degree < n satisfying (45). <u>Hint</u>: Derive a linear system of equations which must be satisfied by the coefficients of P.

The construction of Q is described in (14). It is now easy to
verify that Q interpolates f in n points which generally do not coincide with $t_1,...,t_n$ in (45). We next state an expression for the approximation error.

(48) <u>Lemma</u>. Let $[\alpha,\beta]$ be a closed bounded interval and let
$t_1,...,t_n$ be n points with $\alpha \le t_1 < t_2 < ... < t_n \le \beta$. Further, let f
have n continuous derivatives $f',...,f^{(n)}$ on $[\alpha,\beta]$ and denote by P
the polynomial of degree < n satisfying (45). Then

$$f(t) = P(t) + R(t), \tag{49}$$

where

$$R(t) = \frac{1}{n!} f^{(n)}(\xi) \prod_{i=1}^{n} (t-t_i), \tag{50}$$

with the unknown point ξ lying in a subinterval of $[\alpha,\beta]$ containing
the points t and $t_1,...,t_n$. In general, ξ depends on t.

The proof of this result is given in Dahlquist-Björck (1974), p. 100.
Using (49) and (50) we get

$$|f(t) - P(t)| \le \frac{1}{n!} |f^{(n)}(t)| \cdot \max_{\alpha \le t \le \beta} |\prod_{i=1}^{n} (t-t_i)|. \tag{51}$$

The approximation error is thus bounded by an expression containing a factor which is independent of f. A natural approach is to make this
second factor as small as possible in the uniform norm. We may here directly apply Theorem (25) to determine the appropriate choice of
$t_1,...,t_n$ in (45); namely,

$$t_i = \frac{\alpha+\beta}{2} + \frac{\beta-\alpha}{2} \cos \frac{i-1/2}{n} \pi, \quad i = 1,2,...,n. \tag{52}$$

To select s_i in (46) we argue as follows. We assume that P in (49)
interpolates f at the points $\hat{t}_i$ of (26). Then

$$\prod_{i=1}^{n} (t-\hat{t}_i) = \frac{2(\beta-\alpha)^n}{4^n} T_n(\frac{2t-\alpha-\beta}{\beta-\alpha}).$$

The maximum of the absolute value of this function is assumed at

$$s_i = \frac{\alpha+\beta}{2} + \frac{\beta-\alpha}{2} \cos \frac{(i-1)\pi}{n}, \quad i = 1,\ldots,n+1, \tag{53}$$

and these points are entered into (46).

(54) **Exercise.** Consider again the problem of approximating $f(t) = t^n$ over a closed bounded interval $[\alpha,\beta]$ by a polynomial of degree $< n$. Verify that the two approaches i) and ii) above give the same results, if we select t_i in (45) according to (52) and s_i in (46) according to (53).

(55) **Exercise.** Assume again that t_i in (45) is given by (52) and s_i in (46) by (53). We determine the polynomials P and Q in the form

$$P = \sum_{r=0}^{n-1} c_r q_r, \quad Q = \sum_{r=0}^{n-1} d_r q_r,$$

where

$$q_r(t) = T_r(\frac{2t-\alpha-\beta}{\beta-\alpha}), \quad r = 0,\ldots,n-1.$$

Use the orthogonality relations (33) and (34) to derive expressions for the coefficients c_r and d_r. Show also that the number ε of (15) is given by

$$\frac{1}{n} \sum_{i=1}^{n+1} {}'' (-1)^{i-1} f(s_i),$$

thus obtaining a lower bound for the achievable approximation error.

Chapter IV
Duality Theory

A major topic of this chapter is the derivation of "strong" duality results, i.e. theorems which specify when $v(D) = v(P)$. Another important topic is the existence of solutions to the problems (P) and (D). We shall give two strong duality theorems, namely (9) of §10 and (7) of §11. They can be used to verify strong duality in most linear optimization problems occurring in practice.

§8 is of independent interest since it gives a geometric representation of the dual problem (D) that also is helpful for the understanding of the numerical procedures to be described in Chapter V, VI, and VII.

§8. GEOMETRIC INTERPRETATION OF THE DUAL PROBLEM

At first we introduce the concept of a convex set and the special case of a convex cone. Their elementary properties are discussed. A very simple geometric representation of the dual problem will be given.

(1) <u>Definition</u>. The set $K \subset R^n$ is said to be *convex* if it has the following property: a_1 and a_2 belonging to K implies that the entire line segment between a_1 and a_2 lies in K. This may be written:

$a_1 \in K$, $a_2 \in K \Rightarrow$

$\lambda a_1 + (1-\lambda)a_2 \in K$, $\lambda \in [0,1]$.

By induction on q we easily establish that if K is convex and $a_1,\dots,a_q \in K$, then

$$\sum_{i=1}^{q} \lambda_i a_i \in K$$

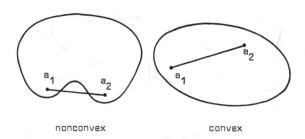

noncovex   convex

Fig. 8.1

if

$$\lambda_1 + \lambda_2 + \ldots + \lambda_q = 1$$

and

$$\lambda_i \geq 0, \quad i = 1,\ldots,q.$$

See Fig. 8.1.

(2) **Definition.** Let A be an arbitrary set of vectors in R^n. We define the *convex hull* of A, denoted Conv (A), to be the set of all vectors $x \in R^n$ admitting a representation

$$x = \sum_{i=1}^{q} \lambda_i a_i \quad (q \geq 1)$$

where

$$\sum_{i=1}^{q} \lambda_i = 1$$

and

$$\lambda_i \geq 0, \quad i = 1,\ldots,q, \quad a_i \in A, \quad i = 1,\ldots,q.$$

Thus the convex hull of A, Conv (A), consists of all possible *convex combinations*

$$x = \sum_{i=1}^{q} \lambda_i a_i, \quad \lambda_i \geq 0, \quad \sum_{i=1}^{q} \lambda_i = 1, \quad q \geq 1 \tag{3}$$

of finitely many vectors from A. The number q can be arbitrarily large. The verification of the following statements is straightforward: Conv (A) is convex for any set A; a convex set which contains A must contain all convex combinations (3); Conv (A) is the *smallest* convex set having A as a subset. See also Fig. 8.2.

Fig. 8.2

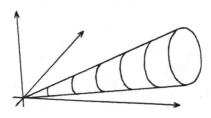

Fig. 8.3

(4) <u>Definition</u>. A *convex cone* is a convex set with the property that if $x \in C$ then $\lambda x \in C$ for all $\lambda \geq 0$. See Fig. 8.3. Let K be a convex set. Then all vectors

$$y = \lambda x \quad \text{where} \quad x \in K, \quad \lambda \geq 0 \tag{5}$$

form a convex cone C which we shall denote Cone (K), the *conic hull* of the convex set K. It is straightforward to verify that Cone (K) is the smallest convex cone containing K.

(6) <u>Definition</u>. Let A be an arbitrary subset of R^n. We shall use the word *convex conic hull* of A and the notation Cone (Conv (A)) for the conic hull of the convex hull of A. Instead of Cone (Conv (A)) we shall sometimes write $CC(A)$.

By (5), we obtain $CC(A)$ by forming all nonnegative multiples of all convex combinations (3) of elements of A.

$$CC(A) = \{z \mid z = \sum_{i=1}^{q} x_i a_i, \quad x_i \geq 0, \quad i = 1,\ldots,q,$$
$$a_i \in A, \quad i = 1,\ldots,q, \quad q \geq 1\}. \tag{7}$$

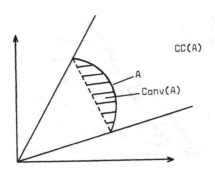

Fig. 8.4

Thus the convex conic hull consists of all nonnegative linear combinations of elements of the set A. We shall apply the concepts introduced above to the set of vectors which occurs by the formulation in §4 of the dual pair (P) - (D). The constraints of the primal problem

$$a(s)^T y \geq b(s), \quad s \in S,$$

can be expressed in terms of the set of vectors

$$A_S = \{a(s) \mid s \in S\} \subset R^n. \tag{8}$$

Combining (16) of §4 and (17) of §4 with (7) we find that $\{s_1, \dots, s_q, x_1, \dots, x_q\}$ is feasible for the dual problem if and only if the vector c may be written as a nonnegative linear combination of the vectors $a(s_1), \dots, a(s_q)$ with coefficients $x_1, \dots, x_q$. Thus (D) has feasible solutions if and only if c lies in $CC(A_S)$.

Since the convex cone $CC(A_S)$ will play a major role in our presentation we shall introduce a special notation.

(9) <u>Definition</u>. The convex conic hull of A_S will be denoted M_n and called the *moment cone* of the optimization problem (P),

$$M_n = CC(A_S).$$

The words "moment cone" are traditional and will not be elaborated upon.

From the remarks preceding the last definition we get the following statement:

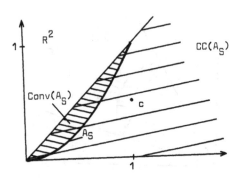

Fig. 8.5

(10) <u>Lemma</u>. The dual problem (D) is feasible if and only if

$c \in M_n$.   (11)

(12) <u>Example</u>. Put $n = 2$, $S = [0,1]$ and

$$a(s) = \begin{pmatrix} s \\ s^2 \end{pmatrix}, \quad s \in [0,1].$$

The sets A_S, Conv (A_S), and $M_n = CC(A_S)$ are indicated in Fig. 8.5.
Consider now the optimization problem:

Minimize $y_1 + \frac{1}{2} y_2$ subject to

(P)

$sy_1 + s^2 y_2 \geq e^s - 1$, $\quad s \in [0,1]$.

Here $c = (1, 1/2)^T$. We see from Fig. 8.5 that c is in M_n and that
the dual (D) of (P) above is feasible. (Exercise: express c as a nonnegative combination of two suitable vectors $a(s_1)$ and $a(s_2)$!).

(13) <u>Exercise</u>. Consider the same example as in (12) but with the
modification

Minimize y_1.

Is the corresponding dual feasible?

We have hitherto permitted an arbitrarily large natural number q
in the representation (7) of the convex hull. However, $CC(A)$ is a subset of a finite-dimensional vector space R^n and one might conjecture

that at most $q = n$ vectors a_i from A are required for the representation of a vector z in $CC(A)$. (Try some simple examples!)

We now prove a general statement to this effect.

(14) <u>Reduction Theorem</u>. Let the vector $z \in R^p$ $(p \geq 1)$ be a non-negative linear combination of the q vectors $z_1,\ldots,z_q$ in R^p $(q \geq 1)$, i.e.

$$z = \sum_{i=1}^{q} x_i z_i, \quad x_i \geq 0, \quad i = 1,\ldots,q. \tag{15}$$

Then z admits a representation

$$z = \sum_{i=1}^{q} \bar{x}_i z_i, \quad \bar{x}_i \geq 0, \quad i = 1,\ldots,q. \tag{16}$$

such that at most p of the numbers $\bar{x}_i$ are nonzero and such that the set of vectors z_i with $\bar{x}_i > 0$, $\{z_i \mid \bar{x}_i > 0\}$, is linearly independent.

<u>Proof</u>: The proof is constructive and we will show how to arrive at a representation (16) from (15) by means of finitely many arithmetic operations. If $z_1,\ldots,z_q$ already are linearly independent then $q \leq p$ and the numbers $x_i = \bar{x}_i$ are uniquely determined by (15) and (16), and there is nothing more to prove. We assume therefore that $z_1,\ldots,z_q$ are linearly dependent. We will demonstrate how to reduce the number of positive terms in (15) step by step until the corresponding vectors become linearly independent. The linear dependence of $z_1,\ldots,z_q$ means that there are numbers $\alpha_1,\ldots,\alpha_q$ such that

$$\sum_{i=1}^{q} \alpha_i z_i = 0. \tag{17}$$

Hence we have for each r with $\alpha_r \neq 0$

$$z_r = -\sum_{i \neq r} \frac{\alpha_i}{\alpha_r} z_i.$$

Entering this relation into (15) we get

$$z = \sum_{\substack{i=1 \\ i \neq r}}^{q} (x_i - x_r \frac{\alpha_i}{\alpha_r}) z_i. \tag{18}$$

Hence we have got a representation of z as a linear combination of $q-1$ of the vectors $z_1,\ldots,z_q$. We must now also show that r can be chosen so that (18) becomes a nonnegative linear combination, i.e.

$$x_i - x_r \frac{\alpha_i}{\alpha_r} \geq 0, \quad i = 1,\ldots,r-1,r+1,\ldots,q. \tag{19}$$

We now select r such that

$$\alpha_r > 0. \tag{20}$$

(If all α_r in (17) are nonpositive we multiply (17) by -1.) Since $x_i \geq 0$ and $\alpha_r > 0$ we conclude that

$$x_i - x_r \frac{\alpha_i}{\alpha_r} \geq 0 \quad \text{if} \quad \alpha_i \leq 0.$$

The condition (19) is thus met when $\alpha_i \leq 0$. We next discuss the case $\alpha_i > 0$. Then (19) implies

$$\frac{x_i}{\alpha_i} \geq \frac{x_r}{\alpha_r} .$$

This condition and consequently (19) is certainly met if we determine r such that

$$\frac{x_r}{\alpha_r} = \min\left\{\frac{x_i}{\alpha_i} \mid \alpha_i > 0\right\}. \tag{21}$$

Then (18) expresses z as a nonnegative linear combination of the $q-1$ vectors $z_1,\ldots,z_{r-1}, z_{r+1},\ldots,z_q$. This procedure may be repeated until we have determined a representation (16) such that those vectors which belong to nonnegative coefficients $\bar{x}_i$ are linearly independent.

(22) <u>Example</u>. ($p = 2$, $q = 4$). Let

$$z_1 = \begin{pmatrix} 1 \\ 1 \end{pmatrix}, \quad z_2 = \begin{pmatrix} 1 \\ 2 \end{pmatrix}, \quad z_3 = \begin{pmatrix} 2 \\ 1 \end{pmatrix}, \quad z_4 = \begin{pmatrix} 2 \\ 2 \end{pmatrix}, \quad z = \begin{pmatrix} 7/4 \\ 2 \end{pmatrix} .$$

Then z admits the representation

$$z = \frac{1}{4} z_1 + \frac{1}{2} z_2 + \frac{1}{4} z_3 + \frac{1}{4} z_4. \tag{23}$$

This relation corresponds to (15). The vectors are linearly dependent, of course. Thus we have, for example,

$$z_1 = \frac{1}{3} z_2 + \frac{1}{3} z_3$$

or

$$-3z_1 + z_2 + z_3 = 0.$$

This corresponds to (17) with $\alpha_1 = -3$, $\alpha_2 = 1$, $\alpha_3 = 1$, $\alpha_4 = 0$. Since we must have $\alpha_r > 0$, $r = 2$ or $r = 3$ meets this condition. By (21) we must next determine the smaller of the quotients

$$\frac{x_2}{\alpha_2} = \frac{1}{2} \quad \text{and} \quad \frac{x_3}{\alpha_3} = \frac{1}{4}.$$

Thus we should take $r = 3$ and (18) gives

$$z = z_1 + \frac{1}{4} z_2 + \frac{1}{4} z_4. \tag{24}$$

Thus z_3 no longer appears in (24), in contrast to the representation (23). Carry out another reduction step, this time on (24), and obtain z as a nonnegative linear combination of two of the vectors z_1, z_2, z_4. Is it possible to carry the reduction even further?

(25) <u>Exercise</u>. Prove the Lemma of Carathéodory: Every vector $z \in \text{Conv}(A) \subset R^n$ admits a representation

$$z = \sum_{i=1}^{n+1} x_i a_i$$

where

$$a_1, \ldots, a_{n+1} \in A, \quad x_1, \ldots, x_{n+1} \geq 0 \quad \text{and} \quad \sum_{i=1}^{n+1} x_i = 1.$$

From the Reduction Theorem (14) we obtain the following result:

(26) <u>Theorem</u>. Let $\{s_1, \ldots, s_q, x_1, \ldots, x_q\}$ with $q \geq 1$ be feasible for (D); i.e.

$$\sum_{i=1}^{q} a_r(s_i)x_i = c_r, \quad r = 1, \ldots, n, \tag{27}$$

and

$$x_i \geq 0, \quad i = 1, \ldots, q.$$

Then there is a subset $\{s_{i_1}, \ldots, s_{i_n}\}$ of $\{s_1, \ldots, s_q\}$ and numbers $\bar{x}_{i_1}, \ldots, \bar{x}_{i_n}$ with the properties

$\{s_{i_1}, \ldots, s_{i_n}, \bar{x}_{i_1}, \ldots, \bar{x}_{i_n}\}$ is also feasible to (D); i.e.

$$\sum_{j=1}^{n} a_r(s_{i_j})\bar{x}_{i_j} = c_r, \quad r = 1, \ldots, n, \tag{28}$$

$$\bar{x}_{i_j} \geq 0, \quad j = 1, \ldots, n.$$

The vectors $a(s_{i_j})$ which belong to *positive* numbers $\bar{x}_{i_j}$ are linearly independent.

In Theorem (26) we have tacitly assumed that S has at least n elements, i.e. at least n constraints occur by (P). As a rule, this requirement is met. (In many applications S has infinitely many elements or S is the result of "sufficiently fine" discretization and $|S|$ becomes very large.)

One can always achieve $|S| \geq n$ by adding the trivial constraint $0^T y \geq 0$ to (P) sufficiently many times. This operation does not change M_n.

However, we *cannot* conclude from Theorem (26) that we only need to consider feasible solutions $\{s_1,\ldots,s_q, x_1,\ldots,x_q\}$ of the dual problem (D) with $q \leq n$ and that we can put $q = n$ from the start in the formulation of the dual problem. It is quite possible that by the transition from (27) to (28) by means of the Reduction Theorem (14) it happens that

$$\sum_{j=1}^{n} b(s_{i_j})\bar{x}_{i_j} < \sum_{i=1}^{q} b(s_i)x_i.$$

If one wants to make sure that the value of the dual preference function does not change, then one must apply the Reduction Theorem on the n+1 *equations*

$$\sum_{i=1}^{q} b(s_i)x_i = c_0,$$

$$\sum_{i=1}^{q} a_r(s_i)x_i = c_r, \quad r = 1,\ldots,n. \tag{29}$$

We obtain then the important result that n+1 points s_{i_j} "are enough". Thus we may put $q = n+1$ from the start in the formulation of (D).

(30) <u>The dual problem (D)</u>.

$$\text{Maximize} \quad \sum_{i=1}^{n+1} b(s_i)x_i$$

subject to the constraints

$$\sum_{i=1}^{n+1} a_r(s_i)x_i = c_r, \quad r = 1,\ldots,n,$$

$$s_1,\ldots,s_{n+1} \in S,$$

$$x_1,\ldots,x_{n+1} \geq 0.$$

We will show in (7) of §12 that if (D) has a solution then we can even put $q = n$ from the outset.

From the preceding argument, in particular (29), we are led to introduce yet another moment cone $M_{n+1} \subset R^{n+1}$. We adjoin the real number $b(s)$ to the vector $a(s)$ and consider the vectors

$$\tilde{a}(s) = \begin{pmatrix} b(s) \\ a_1(s) \\ \vdots \\ a_n(s) \end{pmatrix} = (b(s), a_1(s), \ldots, a_n(s))^T \in R^{n+1}.$$

Then we can write (29) in the form

$$\sum_{i=1}^{q} \tilde{a}(s_i) x_i = (c_0, c_1, \ldots, c_n)^T. \tag{31}$$

Following the pattern of (8), we let

$$\tilde{A}_S = \{\tilde{a}(s) \mid s \in S\} \subset R^{n+1}$$

and can then define M_{n+1} .

(32) <u>Definition.</u> The moment cone M_{n+1} associated with the optimization problem (P) is the convex conic hull of $\tilde{A}_S$;

$$M_{n+1} = CC(\tilde{A}_S).$$

By the definition of the convex conic hull (see (7)) every vector $\tilde{z} \in M_{n+1}$ admits the representation

$$\tilde{z} = \sum_{i=1}^{q} \tilde{a}(s_i) x_i, \quad x_i \geq 0. \tag{33}$$

By comparison with (31) we realize that $\{s_1, \ldots, s_q, x_1, \ldots, x_q\}$ is feasible for (D) with the corresponding value c_0 of the dual preference function if and only if

$$(c_0, c_1, \ldots, c_n)^T \in M_{n+1}. \tag{34}$$

(We may put $q = n+1$ in (31) and (33) by the Reduction Theorem.)

From (34) we obtain a "geometric" formulation of the dual problem. It will be fundamental for the discussion to follow.

(35) <u>A "geometric" formulation of the dual problem.</u>

Maximize c_0

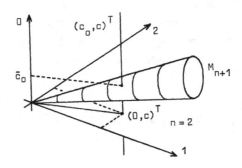

Fig. 8.6

subject to the constraint

$$(c_0,\ldots,c_n)^T \in M_{n+1}.$$

It is, at least in principle, clear how to get from a solution $(\bar{c}_0,c)$ of (35) to a dual solution $\{\bar{s}_1,\ldots,\bar{s}_{n+1}, \bar{x}_1,\ldots,\bar{x}_{n+1}\}$ (and vice versa). Since $(\bar{c}_0,c_1,\ldots,c_n)^T \in M_{n+1}$,

$$\bar{c}_0 = \sum_{i=1}^{n+1} b(\bar{s}_i)\bar{x}_i$$

and

$$c = \sum_{i=1}^{n+1} a(\bar{s}_i)\bar{x}_i,$$

where $\bar{x}_i$ are nonnegative numbers and $\bar{s}_i \in S$, $i = 1,\ldots,n+1$. $\{\bar{s}_1,\ldots,\bar{s}_{n+1}, \bar{x}_1,\ldots,\bar{x}_{n+1}\}$ is hence a solution to (D).

Fig. 8.6 gives a geometric illustration of the dual problem (35). We seek that point $(\bar{c}_0,c)^T$ of the straight line

$$\{(c_0,c_1,\ldots,c_n),\quad c_0 \in R\}$$

which belongs to M_{n+1} and whose first component is as large as possible. We mention also the special case of linear programming. There we find

$$M_{n+1} = \{z \in R^{n+1} \mid z = \sum_{i=1}^{m} \tilde{a}_i x_i = \tilde{A}x,\quad x = (x_1,\ldots,x_m)^T \geq 0\}. \quad (36)$$

Let

$$\tilde{A} = \begin{pmatrix} b_1 & b_2 & \dots & b_m \\ a_{11} & a_{12} & \dots & a_{1m} \\ a_{21} & a_{22} & \dots & a_{2m} \\ \vdots & \vdots & & \\ a_{n1} & a_{n2} & \dots & a_{nm} \\ \uparrow & \uparrow & & \uparrow \\ a_1 & a_2 & & a_m \end{pmatrix}$$

We can now write the condition (34) in the form

$$c_0 = b^T x,$$
$$c = Ax,$$
$$x \geq 0,$$

The dual problem (LD) is then equivalent to the following problem (we write x_0 for c_0)

$$\text{Minimize } x_0 \text{ subject to } -x_0 + b^T x = 0,$$
$$Ax = c,$$
$$(x_1, \dots, x_m)^T \geq 0.$$

§9. SOLVABILITY OF THE DUAL PROBLEM

The following important theorem on the solvability of the dual problem (D) is an immediate consequence of the formulation (35) of §8.

(1) **Theorem**. Let a given linear optimization problem be such that M_{n+1} is closed and the dual problem (D) is bounded (i.e. it is in state "B" - see Diagram (1), §5). Then problem (D) has a solution.

Proof: We note that $v(D)$ is the maximum of the continuous function f given by

$$f(z_0, \dots, z_n) = z_0,$$

defined on the closed and bounded set

$$M_{n+1} \cap \{(z_0, z)^T \mid v(D) - i \leq z_0 \leq v(D), \ z = c\}.$$

Here $c \in R^n$ is the vector appearing in the preference function of (P).

Theorem (1) is very useful since there are simple criteria for ascertaining that M_{n+1} is closed. They are applicable for important classes of linear optimization problems. We shall show in §10 that the dual pair (P) - (D) has no duality gap under the assumptions of Theorem (1).

Quite often we shall encounter a special class of problems where the index set S and the functions $a_1, \ldots, a_n$, b which appear in the constraints of (P),

$$\sum_{r=1}^{n} a_r(s)y_r \geq b(x), \quad s \in S,$$

satisfy the following assumptions:

(2) <u>General assumptions on (P)</u>. S is a compact subset of R^k and the real-valued functions $a_1, \ldots, a_n, b$ which are defined on S are continuous there.

This assumption is valid for the examples (3) of §3 and (7) of §4 (with $k = 1$) but not for (4) of §3. For the special case of linear programming (2) holds trivially; since S is finite *every* real-valued function on S is continuous. We can then always assume that

$$S = \{1, \ldots, m\} \subset R.$$

(3) <u>Definition</u>. If there is a vector $\tilde{y} = (\tilde{y}_1, \ldots, \tilde{y}_m)^T \in R^n$ such that

$$\sum_{r=1}^{n} a_r(s)\tilde{y}_r > b(s), \quad s \in S, \tag{4}$$

then (P) is said to meet the Slater condition. If (P) satisfies (4) then we also call (P) *superconsistent* since (4) is a sharpening of the statement that $\tilde{y}$ is feasible for (P).

Suppose now that Assumption (2) is satisfied. Then the Slater condition (4) is met if one of the functions $a_1, \ldots, a_n$ is constant, e.g.

$$a_1(s) = 1, \quad s \in S.$$

Indeed, (4) is met if we take

$$\tilde{y} = (\tilde{y}_1, 0, \ldots, 0)^T.$$

where

$$\tilde{y}_1 > \max_{s \in S} b(s).$$

This is possible since b is continuous on a compact set. (Compare (13) of §2.)

(5) <u>Remark</u>. The Slater condition is an example of the so-called *regularity* conditions which are introduced in the theory of optimization and which play a major role in the derivation of theorems on duality and on existence of solutions. We shall encounter another regularity condition in §11.

(6) <u>Exercise</u>. Consider (P) given Assumption (2) and the Slater condition (4). Show that the set of vectors feasible to (P) has interior
points. <u>Hint</u>: Let $\tilde{y} \in R^n$ satisfy (4). Show that there is an $\varepsilon > 0$
such that all vectors y with $|y-\tilde{y}| \le \varepsilon$ are feasible for (4).

The two theorems to follow can be used to establish the existence of
solutions of the dual of most linear optimization problems encountered in
practice.

(7) <u>Theorem</u>. Suppose that Assumption (2) is satisfied and that (P)
meets the Slater condition. Then the moment cone M_{n+1} is closed.

In order to carry out the proof of this theorem we need the following result which is of independent interest.

(8) <u>Lemma</u>. Let $A \subset R^p$ be a compact set. Then its convex hull,
Conv (A), is also compact.

<u>Proof</u>: By (25) of §8, Conv (A) is generated by means of all possible
linear combinations

$$\sum_{i=1}^{p+1} a_i x_i$$

where

$$a_1, \ldots, a_{p+1} \in A \quad \text{and} \quad (x_1, \ldots, x_{p+1}) \in D,$$

where the set $D \subset R^{p+1}$ is defined by

$$D = \{x \in R^{p+1} \mid x_i \ge 0, \ i = 1, \ldots, p+1, \ \text{and} \ \sum_{i=1}^{p+1} x_i = 1\}.$$

Hence Conv (A) is the image of the compact set

$$A \times A \times \ldots \times A \times D$$
$$\text{(p+1 times)}$$

under the continuous mapping

$$(a_1, \ldots, a_{p+1}, \ x_1, \ldots, x_{p+1}) \rightarrow \sum_{i=1}^{p+1} a_i x_i.$$

Since A was compact, Conv (A) must be compact as well. (See the remark
after (12) of §2.)

<u>Proof of Theorem (7)</u>: Let z be an arbitrary vector in $\bar{M}_{n+1}$. We
will show that then z must be in M_{n+1} also. By Definition (32) of §8,

$$M_{n+1} = CC(\tilde{A}_S).$$

Thus to $z \in \bar{M}_{n+1}$ we may associate a sequence $\{h_i\}_{i>1}$ in $\text{Conv}(\tilde{A}_S)$ and a sequence of nonnegative numbers $\{\lambda_i\}_{i>1}$ such that

$$z = \lim_{i \to \infty} \lambda_i h_i. \tag{9}$$

The set $\tilde{A}_S$ is compact since S is compact and $a_1, \ldots, a_n$, b are continuous. By Lemma (8), $\text{Conv}(\tilde{A}_S)$ is compact. We may therefore pick a subsequence of $\{h_i\}_{i>1}$ which converges to a vector $h \in \text{Conv}(\tilde{A}_S)$. Thus we may as well assume from the outset that the sequence $\{h_i\}_{i>1}$ in (9) is such that

$$\lim_{i \to \infty} h_i = h, \quad h \in \text{Conv}(\tilde{A}_S).$$

If now the sequence $\{\lambda_i\}_{i>1}$ is *bounded* we can in the same way assume that it converges to $\lambda > 0$. Then we obtain

$$z = \lim_{i \to \infty} \lambda_i h_i = \lim_{i \to \infty} \lambda_i \lim_{i \to \infty} h_i = \lambda h$$

and from $h \in \text{Conv}(\tilde{A}_S)$, $\lambda \geq 0$, it follows that $z = \lambda h \in CC(\tilde{A}_S) = M_{n+1}$ as was to be established. We next consider the remaining case when $\{\lambda_i\}_{i>1}$ is *unbounded*. Then we may assume, if necessary by using a suitable subsequence, that

$$\lambda_i > 0, \quad i = 1, 2, \ldots,$$

and

$$\lim_{i \to \infty} 1/\lambda_i = 0.$$

Thus we get

$$h = \lim_{i \to \infty} h_i = \lim_{i \to \infty} \frac{1}{\lambda_i} \lambda_i h_i = \lim_{i \to \infty} \frac{1}{\lambda_i} \lim_{i \to \infty} \lambda_i h_i = 0z = 0.$$

This means that the null vector of R^{n+1} lies in $\text{Conv}(\tilde{A}_S)$. Hence there are $q \geq 1$ nonnegative numbers $\alpha_1, \ldots, \alpha_q$ and q points $s_1, \ldots, s_q$ in S such that

$$0 = \sum_{i=1}^{q} \tilde{a}(s_i)\alpha_i$$

and

$$\sum_{i=1}^{q} \alpha_i = 1. \tag{10}$$

From the definition of $\tilde{a}(s)$ (see (30)- (31) of §8) this implies that

$$0 = \sum_{i=1}^{q} b(s_i)\alpha_i$$

and

$$0 = \sum_{i=1}^{q} a_r(s_i)\alpha_i, \quad r = 1,\ldots,n.$$

Let $y \in R^n$ be an arbitrary vector. The last two equations now give

$$0 = \sum_{i=1}^{q} \alpha_i \left(\sum_{r=1}^{n} y_r a_r(s_i) - b(s_i) \right) \tag{11}$$

Since problem (P) is required to meet the Slater condition there is a $\tilde{y} \in R^n$ such that

$$\sum_{r=1}^{n} \tilde{y}_r a_r(s_i) - b(s_i) > 0, \quad i = 1,\ldots,q.$$

If we now put $y = \tilde{y}$ in (11) we get, since $\alpha_i \geq 0$, that $\alpha_1 = \alpha_2 = \ldots = \alpha_q = 0$ must hold, contradicting (10). This rules out the possibility that $\{\lambda_i\}_{i>1}$ is unbounded. Hence we have established the theorem.

(12) <u>Example</u>. Consider the constraint

$$s^2 y_1 \geq s, \quad s \in [0,1].$$

Here we have $n = 1$, $S = [0,1]$, $a_1(s) = s^2$, $b(s) = s$. The Slater condition is not met since $a_1(0) = b(0) = 0$. M_{n+1} is not closed since the vectors $(x_1,0)^T$, $x_1 > 0$ are in $\bar{M}_{n+1}$ but not in M_{n+1}.

(13) <u>Exercise</u>. Consider the problem of uniform approximation over a compact set, discussed in §6. Show that the Slater condition is met. In §4 we showed that if (P) and (D) are consistent, then (D) has a finite value. Combining (1) and (7) we get the following statement on the existence of solutions to (D).

(14) <u>Theorem</u>. Let the dual pair (P) - (D) have the properties

i)   Assumption (2) is satisfied,
ii) (D) is feasible,
iii) (P) meets the Slater condition.

Then (D) is solvable.

This theorem will be sharpened significantly in 12 of §10.

We now treat linear programming and show that the corresponding moment cone M_{n+1} as defined in (36) of §8 is closed in this case.

We shall say that cones of the form

$$C = \{z \in R^p \mid z = Ax, \quad (x_1, \ldots, x_m) \geq 0\}$$

are *finitely generated*. In the case of linear programming, M_{n+1} is finitely generated, and the following theorem establishes that M_{n+1} is closed.

(15) **Theorem.** Every finitely generated cone in R^p is closed.

Proof: We consider first the case when the $p \times m$ matrix A has rank p. Then the rows of A are linearly independent. Let now $\{z^j\}_{j>1}$ be a convergent sequence in C such that

$$z^j \to \bar{z}. \tag{16}$$

We want to show that $\bar{z}$ is also in C. Every z^j can be written as a nonnegative linear combination of at most p linearly independent column vectors of A, by the Reduction Theorem (14) of §8. We may now, for each j, supplement this set of column vectors by picking suitable column vectors from the remaining ones to get a basis for R^p. Then there is for each $j \geq 1$ an index set $I_j \subset \{1, \ldots, m\}$ containing p elements and a vector $x^j \to R^p$ such that

$$z^j = A_j x^j, \quad x^j \geq 0.$$

Here A_j is formed of the columns from A corresponding to I_j. Thus

$$x^j = A_j^{-1} z^j, \quad j \geq 1.$$

However, there are only finitely many matrices A_j. Hence there is among these a fixed matrix $\bar{A}$ and a subsequence $\{j(k)\}_{k\geq 1}$ of natural numbers such that

$$x^{j(k)} = \bar{A}^{-1} z^{j(k)}, \quad k \geq 1.$$

Hence we get from (16)

$$x^{j(k)} \to \bar{x} = \bar{A}^{-1} \bar{z}.$$

Since $x^{j(k)} \geq 0$ we must have $\bar{x} \geq 0$. The relation

$$\bar{z} = \bar{A} \bar{x}$$

then implies that $\bar{z} \in C$ which was the desired conclusion. We now treat the remaining case when the rank of A is less than p. We may assume that the rows of A are ordered such that the first p^1 rows are linearly independent $(1 \leq p^1 < p)$ and the remaining rows are linear combinations of the first p^1 ones. (We have, of course, excluded the trivial case $A = 0$ from further consideration.) Then every $z \in C$ may be written

$$z = (z_1, z_2)^T, \quad z_1 \in R^{p^1}, \quad z_2 \in R^{p-p^1},$$

where

$$z_1 = \hat{A}x, \quad x \in R^m, \quad x \geq 0, \tag{17}$$

and

$$z_2 = Bz_1. \tag{18}$$

Here $\hat{A}$ is a $p^1 \times m$ matrix and B a $(p-p^1) \times p^1$ matrix. We next define the cone $\tilde{C}$ associated with (17) and argue as above and use (18) to arrive at the desired result

$$z = \begin{pmatrix} \hat{A} \\ B\hat{A} \end{pmatrix} x.$$

Combining Theorems (15) and (1) we conclude that (LP) is solvable when (LD) is bounded. We saw in (37) of §4 that every problem in the form of (LP) may be transformed into an equivalent problem in the form of (LD). Hence a corresponding existence theorem is valid for (LP) as well. This fact we summarize in the

(19) **Theorem.** Consider the dual pair (LP) - (LD) of linear programs. If both of these problems are consistent then they both have solutions.

In the next section we shall also show that no duality gap can occur under the assumptions of Theorem (19).

§10. SEPARATION THEOREM AND DUALITY

We shall start this section by developing a fundamental tool to be used in the proof of strong duality theorems, namely the statement that a point outside a closed convex set in R^p may be "separated" from this set by a hyperplane in the sense of the following definition.

(1) **Definition.** Let M be a nonempty, closed and convex subset of R^p and $z \notin M$ a fixed point. The hyperplane

$$H(y; \eta) = \{x \in R^p \mid y^T x = \eta\}$$

is said to *separate* z from M if

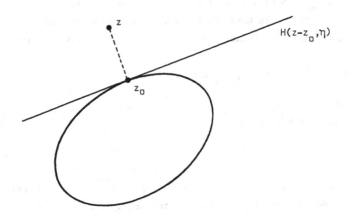

Fig. 10.1. Separating hyperplane

$$y^T x \leq \eta < y^T z, \quad x \in M.$$

From geometric considerations (see Fig. 10.1) one is led to believe that a vector y which defines a separating hyperplane is obtained by determining the projection z_0 of z on M and putting $y = z - z_0$. This will turn out to be the correct procedure. We will therefore first show the existence of a unique projection point. (See (4).)

To give a motivation for the argument to follow we shall first indicate the fundamental role of the concept of separating hyperplanes in the theory of the dual pair (P) - (D).

Assume that the hyperplane

$$H(y;0) = \{z \in R^{n+1} \mid \sum_{r=0}^{n} z_r y_r = 0\}$$

separates the moment cone M_{n+1} from the point $v \notin M_{n+1}$. Thus all of M_{n+1} lies on one side of the hyperplane. Hence

$$0 \geq \sum_{r=0}^{n} z_r y_r, \quad \text{all} \quad (z_0, \ldots, z_n) \in M_{n+1}. \tag{2}$$

In particular, since $M_{n+1} = CC(\tilde{A}_S)$ we have

$$z = \tilde{a}(s) = (b(s), a_1(s), \ldots, a_n(s))^T \in M_{n+1}$$

for all $s \in S$. Thus we find from (2) that

$$0 \geq b(s)y_0 + \sum_{r=1}^{n} a_r(s)y_r, \quad s \in S.$$

If $y_0 > 0$ holds, then the last relation takes the form

$$\sum_{r=1}^{n} a_r(s) \frac{-y_r}{y_0} \geq b(s), \quad s \in S.$$

Hence the vector $y = (-y_1/y_0,\ldots,-y_n/y_0)$ is feasible for (P).

(3) <u>Exercise</u>. Let y be feasible for (P). Give a hyperplane passing through the origin such that M_{n+1} is on one side of this hyperplane.

(4) <u>Projection Theorem</u>. Let $M \subset R^p$ be a nonempty, closed, convex set and let z be a fixed point outside of M. Then there is exactly one vector $z_0 \in M$ which lies "closest" to z. That is, z_0 is such that

$$0 < |z - z_0| \leq |z - x|, \quad \text{all} \quad x \in M.$$

<u>Proof</u>: Since M is closed and $z \notin M$ we find

$$\rho = \inf_{x \in M} |z - x| > 0.$$

Obviously, it is sufficient to search for the vector z_0 in the set

$$\tilde{M} = M \cap \{x \in R^p \mid |z - x| \leq 2\rho\}.$$

Now the continuous real-valued function $x \to |z - x|$ assumes its minimum value on the bounded and closed set $\tilde{M}$. Hence there is a $z_0 \in M$ such that

$$|z - z_0| \leq |z - x|, \quad x \in \tilde{M}. \tag{5}$$

From the construction of $\tilde{M}$, (5) holds for all $x \in M$. We must now establish the *uniqueness* of the projection point z_0. Assume therefore that there is a vector $z_1 \neq z_0$ such that

$$|z - z_1| \leq |z - x|, \quad \text{all} \quad x \in M.$$

We now put $z_2 = (z_1 + z_0)/2$. The parallelogram law from (10) of §2 gives

$$|z-z_2|^2 = \frac{1}{4}|(z-z_0) + (z-z_1)|^2 < \frac{1}{4}|(z-z_0) + (z-z_1)|^2$$

$$+ \frac{1}{4}|z_0-z_1|^2 = \frac{1}{4}|(z-z_0) + (z-z_1)|^2 + \frac{1}{4}|(z-z_0) - (z-z_1)|^2$$

$$= \frac{1}{2}(|z-z_0|^2 + |z-z_1|^2) = |z-z_0|^2.$$

implying $|z-z_2| < |z-z_0|$. This contradicts the construction of z_0, hence uniqueness is established.

(6) <u>Separation Theorem</u>. Let $M \subset R^p$ be a nonempty, closed, convex set. Let $z \notin M$ be a fixed point whose projection on M is z_0. If we put $y = z-z_0$ and $\eta = (z-z_0)^T z_0$ we get

$$y^T x \le \eta < y^T z, \quad x \in M; \tag{7}$$

i.e. the hyperplane $H(y;\eta)$ separates z from M .

<u>Proof</u>: Let $x \in M$ be an arbitrary vector and $0 < \eta \le 1$ be a fixed number. Then

$$(1-\mu)z_0 + \mu x = z_0 + \mu(x-z_0) \in M.$$

We also find that

$$|z-z_0|^2 \le |z - (z_0 + \mu(x-z_0))|^2$$
$$= |z-z_0|^2 - 2\mu(z-z_0)^T(x-z_0) + \mu^2 |x-z_0|^2,$$

giving

$$(z-z_0)^T(x-z_0) \le \frac{1}{2} \mu |x-z_0|^2.$$

Letting $\mu \to 0$ we arrive at

$$(z-z_0)^T(x-z_0) \le 0,$$

establishing the leftmost inequality in (7). The other inequality results from the relation

$$0 < |z-z_0|^2 = (z-z_0)^T(z-z_0) = y^T z - y^T z_0 = y^T z - \eta,$$

concluding the proof.

Suppose now that the assumptions of Theorem (6) hold, but specialize M to be a convex cone. Then $x \in M$ implies that $\lambda x \in M$ for all $\lambda > 0$. From (7) we then get

$$y^T(\lambda x) \le \eta, \quad \lambda > 0,$$

or

$$y^T x \le \eta/\lambda, \quad \lambda > 0.$$

Letting $\lambda \to \infty$ we conclude

$$y^T x \le 0, \quad x \in M.$$

Thus if M is a convex cone we may put $\eta = 0$ from the start and write (7) in the form

$$y^T x \leq 0 < y^T z, \quad x \in M. \tag{8}$$

Now we can use the Separation Theorem to establish the duality result which was promised earlier.

(9) <u>First Duality Theorem</u>. Consider the dual pair (P) - (D) and make the following assumptions:

    i)   The dual problem is consistent and has a finite value $v(D)$;

    ii) The moment cone M_{n+1} is closed.

Then (P) is consistent as well and

    $v(P) = v(D)$;

i.e. there is no duality gap. Moreover, (D) is solvable.

<u>Proof</u>: We have already shown that (D) is solvable (Theorem (1) of §9). Thus we have

$$(c_0, c_1, \ldots, c_n)^T \in M_{n+1},$$

but

$$(c_0 + \varepsilon, c_1, \ldots, c_n) \notin M_{n+1}$$

for any $\varepsilon > 0$. Since M_{n+1} is closed we may invoke the Separation Theorem (6) and conclude that there is a hyperplane in R^{n+1} which separates $(c_0 + \varepsilon, c)^T$ from the convex cone M_{n+1} (see (8)). Hence there is a vector $(y_0, y_1, \ldots, y_n)^T \in R^{n+1}$, different from 0, such that

$$\sum_{r=0}^{n} x_r y_r \leq 0 < y_0(c_0 + \varepsilon) + \sum_{r=1}^{n} c_r y_r,$$

$$(x_0, x_1, \ldots, x_n)^T \in M_{n+1}. \tag{10}$$

In (10) we now put

$$(x_0, x_1, \ldots, x_n)^T = (c_0, c_1, \ldots, c_n)^T \in M_{n+1}$$

and obtain

$$y_0 \varepsilon > 0.$$

Since $\varepsilon > 0$ we must hence have $y_0 > 0$. If we now set

$$(x_0, x_1, \ldots, x_n)^T = (b(s), a_1(s), \ldots, a_n(s))^T \in \tilde{A}_s \subset M_{n+1},$$

($s \in S$ is arbitrary) we find from the leftmost inequality in (10) the relation

$$\sum_{r=1}^{n} a_r(s)(-y_r/y_0) \geq b(s), \quad s \in S.$$

Hence the vector

$$\tilde{y} = (-y_1/y_0, -y_2/y_0, \ldots, -y_n/y_0) \in R^n$$

is feasible for (P). The right inequality in (10) implies

$$\sum_{r=1}^{n} c_r(-y_r/y_0) < c_0 + \varepsilon.$$

We now arrive at the following chain of inequalities:

$$v(P) \leq \sum_{r=1}^{n} c_r \tilde{y}_r < c_0 + \varepsilon = v(D) + \varepsilon \leq v(P) + \varepsilon.$$

The first inequality follows from the fact that $\tilde{y}$ is feasible for (P) and the last is a consequence of the weak duality theorem (18) of §4. Thus

$$v(P) - \varepsilon \leq v(D) \leq v(P)$$

for any $\varepsilon > 0$, proving the theorem.

(11) <u>Exercise</u>. Consider again Example (8) of §5. Draw a picture of the moment cone M_{n+1} in R^3 and show that this cone is not closed. <u>Hint</u>: The ray $\{(0,\lambda,0)^T \mid \lambda > 0\}$ lies in $\bar{M}_{n+1}$ but not in M_{n+1}.

In many applications the General Assumption of (2) of §9 is met: S is a compact subset of R^k and the functions $a_1, \ldots, a_n$ and b are continuous on S. We combine the Theorems (7) and (14) of §9 with (9) and arrive at the following useful result:

(12) <u>Theorem</u>. Consider the dual pair (P) - (D) and make the assumptions

i)   General Assumption (2) of §9;

ii) (D) is consistent;

iii) (P) meets the Slater condition.

Then (D) is solvable and the values of (P) and (D) coincide.

We discuss also the case of linear programming, i.e. the dual pair

(LP)   Minimize $c^T y$, $A^T y \geq b$

(LD)   Maximize $b^T x$, $Ax = c$, $x \geq 0$.

Theorem (9) and Theorem (19) of §9 deliver the entire duality theory of
linear programming. We have by Theorem (9) that if (LD) is consistent
and bounded then (LP) is consistent also and the values of (LD) and (LP)
coincide. Using the transformations (37) of §4 we may also conclude that
if (LP) is consistent and bounded then (LD) is consistent as well and the
values of the two problems coincide. From this argument we obtain the
following state and defect diagrams for linear programs. (Compare also
with (1) of §5 and (7) of §5.)

(13) State and defect diagrams for linear programming.

(LD) \ (LP)	IC	B	UB		(LD) \ (LP)	IC	B	UB
IC	1		4		IC	$+\infty$		0
B		5			B		0	
UB	6				UB	0		

State diagram   Defect diagram

(14) Duality theorem for linear programming.

i) A dual pair (LP) - (LD) is in one and only one of the states
    1, 4, 5, 6 of the state diagram (13). All states are realized.
ii) If both programs are consistent (i.e. if state 5 is realized)
    then both problems are solvable and no duality gap occurs.

The reader should construct simple examples ($n = 1$ or $n = 2$) to
show that all the states 1, 4, 5, 6 can be realized.

We recall once more that the First Duality Theorem (9) plays a fundamental role for the argument of this Section. Under the assumptions of
this theorem we may conclude that $v(D) = v(P)$ as well as that (D) has a
solution. However, the assumptions i) and ii) of Theorem (9) do *not*
imply the solvability of (P), as is illustrated by the example in Exercise (13) of §3.

(15) Exercise. Consider the problem of uniform approximation of
(1) of §6. Show that $v(DA) = v(PA)$ and that the dual problem is solvable.

(16) Exercise. We replace the dual (D) by the "modified dual" (D')
as follows:

(D') Maximize c_0 when $(c_0, c)^T \in \bar{M}_{n+1}$.

(Compare with (35) of §8.) Show that the weak duality inequality $v(D') \le$
$v(P)$ is valid for the modified dual pair (P) - (D'). Show also that when

v(D') is finite then (D') is always solvable and that we always have
v(P) = v(D'). (The modified problem (D') is of theoretical interest only.)

(17) Exercise. Use the Separation Theorem (6) to show Farkas' Lemma:
Let $A \subset R^p$ be a nonempty set and $c \in R^n$ a fixed vector. Then
$c \in \overline{CC(A)}$ if and only if all vectors y such that $a^T y \geq 0$ for all
$a \in A$ also satisfy $c^T y \geq 0$. Specialize to the case when A has finitely
many elements.

(18) Remark. The duality theorem (12) can be sharpened somewhat.
(A corresponding statement is true for the First Duality Theorem.) One
can show that the assertions of (12) remain true if we replace the assumption (ii) by

ii') v(P) is finite.

It is easy to establish that ii) and iii) imply ii'). A proof of this
sharpened version of (12) is to be found in Glashoff (1979). For easy
reference we sum up the result, which is quite useful for many applications.

(19) Theorem. Consider the dual pair (P) - (D). Make the following assumptions:

   i)   General assumption (2) of §9 ;
  ii)   v(P) is finite;
 iii)   (P) meets the Slater condition.

Then (D) is solvable and the value of (P) and (D) coincide.

§11. SUPPORTING HYPERPLANES AND DUALITY

In this section we shall prove a theorem which could be said to be a
kind of "dual" to Theorem (9) of §10: from the consistency and boundedness of (D) follows the strong duality result $v(P) = v(D)$ as well as
the solvability of (P) provided certain regularity conditions are met.

For this purpose we will need a corollary to the Separation Theorem
(6) of §10 which states that a supporting hyperplane passes through each
boundary point of a convex set. (See Fig. 11.1.).

(1) Definition. Let M be a nonempty convex subset of R^p and
let $z \in M$ be a fixed point. The hyperplane

$$H(y;\eta) = \{x \in R^p \mid y^T x = \eta\}$$

is said to be a *supporting hyperplane* to M at z if

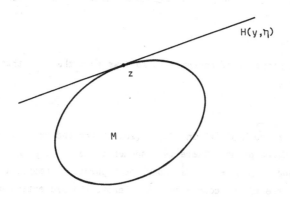

Fig. 11.1 Supporting hyperplane

$$y^T x \leq \eta = y^T z, \quad x \in M.$$

The first point on our agenda is to establish a negative statement whose converse will give us the important theorem on the existence of supporting hyperplanes to boundary points of M.

(2) **Lemma.** Let z be in $\overset{\circ}{M}$, the interior of M. Then there are *no* supporting hyperplanes to M at z.

Proof: Assume M has a supporting hyperplane $H(y;\eta)$ at z. Since $z \in \overset{\circ}{M}$ there is a $\lambda > 0$ such that

$$z_\lambda = z + \lambda y \in M.$$

We find that

$$y^T z_\lambda = y^T z + \lambda y^T y \leq \eta = y^T z,$$

hence

$$\lambda y^T y \leq 0,$$

which contradicts $\lambda > 0$ and $y^T y > 0$. Thus we reach the desired conclusion.

(3) **Theorem.** Let M be a nonempty convex subset of R^p and let z be on the boundary of M ($z \in \text{bd } M = \bar{M} \setminus \overset{\circ}{M}$). Then there is a supporting hyperplane to M at z.

Proof: For every nonempty convex subset $M \subset R^p$ the following statement holds:

$$\text{bd } M = \text{bd } \bar{M}.$$

This elementary property of convex sets follows from the fact that $\text{bd } M = \bar{M} \setminus \overset{\circ}{M}$ since

$$\overset{\circ}{\bar{M}} = \overset{\circ}{M}. \tag{4}$$

We shall show the truth of (4) in (22) - (26) at the end of §11. Now let $z \in \text{bd } M$ be a fixed point. There is a sequence $\{z_i\}$ of points such that $z_i \notin \bar{M}$ and $\lim z_i = z$. We apply the Separation Theorem to the points z_i and the closed convex set $\bar{M}$. Denote the projection of z_i on $\bar{M}$ by z_{io}. Putting $y_i = z_i - z_{io}$ we get

$$y_i^T x < y_i^T z_i, \quad x \in \bar{M}, \quad i = 1,2,\dots \; .$$

Since $z_i \notin \bar{M}$, $y_i \neq 0$, $i = 1,\dots,$ setting

$$\tilde{y}_i = y_i / |y_i|, \quad i = 1,2,\dots,$$

we get

$$|\tilde{y}_i| = 1$$

and

$$\tilde{y}_i^T x < \tilde{y}_i^T z_i, \quad x \in M, \quad i = 1,2,\dots \; . \tag{5}$$

Consider the set

$$B = \{y \in R^p \mid |y| = 1\}.$$

B is closed and bounded, hence compact. Therefore there is a subsequence of $\{\tilde{y}_i\}_{i=1}$ which converges to a point $\tilde{y} \in B$. Applying (5) to this subsequence and passing to the limit we get

$$\tilde{y}^T x \leq \tilde{y}^T z, \quad x \in M,$$

which proves the assertion of the theorem since $\tilde{y} \in B$ and hence $\tilde{y} \neq 0$.

(6) Definition. The dual problem (D) is termed *superconsistent* if

$$c \in \overset{\circ}{M}_n.$$

(7) Second Duality Theorem. Consider the dual pair (P) - (D). Make the assumptions

  i) $v(D)$ is finite;

  ii) (D) is superconsistent, i.e. $c \in \overset{\circ}{M}_n$.

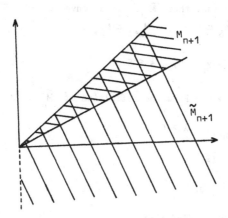

Fig. 11.2. The cones M_{n+1} and $\tilde{M}_{n+1}$

Then (P) is solvable and $v(P) = v(D)$.

Proof: Both (P) and (D) are feasible. Hence the values $v(P)$ and $v(D)$ are finite due to the weak duality lemma. We set as usual

$$\hat{c}_0 = v(D). \tag{8}$$

The vector $(\hat{c}_0, c_1, \ldots, c_n)^T$ lies on the boundary of M_{n+1}. (Otherwise we could find a vector $(c_0, c_1, \ldots, c_n)^T$ with $c_0 > \hat{c}_0$ but still feasible to (D), a fact which would contradict (8).) For the purpose of carrying out the proof we now introduce the following convex cone (see also Fig. 11.2):

$$\tilde{M}_{n+1} = \{(\tilde{z}_0, \tilde{z}_1, \ldots, \tilde{z}_n)^T \mid \text{ there is } (z_0, z_1, \ldots, z_n)^T \in M_{n+1}$$

$$\text{such that } \tilde{z}_0 \leq z_0, \ \tilde{z}_1 = z_1, \ldots, \tilde{z}_n = z_n\}.$$

We find at once that

$$(\hat{c}_0, c_1, \ldots, c_n)^T \in \text{bd } \tilde{M}_{n+1}.$$

By (3) there is a nontrivial supporting hyerplane to $\tilde{M}_{n+1}$ at $(\hat{c}_0, c_1, \ldots, c_n)^T$; i.e. there is a vector $\bar{y} = (y_0, y)^T = (y_0, y_1, \ldots, y_n)^T \neq 0$ such that

$$\bar{y}^T z \leq 0 = y_0 \hat{c}_0 + y^T c, \quad z \in \tilde{M}_{n+1}. \tag{9}$$

We have used here the fact that $\tilde{M}_{n+1}$ is a convex cone. (See (8) of §10.)
(9) implies, since $\tilde{A}_S \subset CC(A_S) = M_{n+1} \subset \tilde{M}_{n+1}$,

$$y_0 b(s) + \sum_{r=1}^{n} a_r(s) y_r \leq 0, \quad s \in S. \tag{10}$$

We now show that $y_0 > 0$. From the definition of $\tilde{M}_{n+1}$ it follows that

$$(\hat{c}_0 - \lambda, c)^T \in \tilde{M}_{n+1}, \quad \lambda > 0.$$

We therefore get from (9)

$$y_0 \hat{c}_0 - y_0 \lambda + y^T c \leq 0.$$

Since $y_0 \hat{c}_0 + y^T c = 0$ we find that

$$-y_0 \lambda \leq 0, \quad \lambda > 0,$$

and hence $y_0 \geq 0$. We must now rule out the possibility $y_0 = 0$.
     Putting $y_0 = 0$, we get from (9) that

$$\sum_{r=1}^{n} y_r z_r \leq \sum_{r=1}^{n} c_r y_r, \quad z \in M_n. \tag{11}$$

M_n is the projection of M_{n+1} on the subspace of R^{n+1} defined through
the condition $z_0 = 0$. Therefore, (11) means that there is a nontrivial
supporting hyperplane to M_n at c. (Since $\bar{y} \neq 0$ and $y_0 = 0$ we must
have $(y_1,\ldots,y_n) \neq 0$.) But this contradicts the fact that $c \in \overset{\circ}{M}_n$
(Lemma (2)). Hence we have established that $y_0 > 0$. We now let

$$\tilde{y}_r = -y_r/y_0, \quad r = 1,\ldots,n,$$

and obtain, from (10),

$$\sum_{r=1}^{n} a_r(s) \tilde{y}_r \geq b(s), \quad s \in S.$$

Thus $(\tilde{y}_1,\ldots,\tilde{y}_n)^T$ is feasible for (P) and hence

$$v(D) \leq v(P) \leq \sum_{r=1}^{n} c_r \tilde{y}_r.$$

By (9) we conclude that

$$\sum_{r=1}^{n} c_r \tilde{y}_r = \hat{c}_0 = v(D).$$

Hence we have shown that $v(P) = v(D)$ and $(\tilde{y}_1, \ldots, \tilde{y}_n)^T$ solves the problem (P).

The Second Duality Theorem just established can be applied to the problem of uniform approximation defined in (1) of §6. We obtain immediately (without requiring the set T to be compact or the functions $a_1, \ldots, a_n$ and b to be continuous) that

$$v(DA) = v(PA)$$

(strong duality) and that the primal problem has a solution (see also (15) of §10):

(12) **Theorem.** Consider the approximation problem (PA) of (1) of §6. Let $v_1, \ldots, v_n$ be linearly independent on T; i.e.

$$\sum_{r=1}^{n} y_r v_r(t) = 0, \quad t \in T$$

implies $y_1 = y_2 = \ldots = y_n = 0$. Then (PA) is solvable and the values of (PA) and (DA) coincide.

Proof: We will show that the linear optimization problem which is equivalent to (PA) satisfies the assumptions of (7). We must verify that the vector $c = (0, \ldots, 0, 1)^T$ of (2) of §6 lies in the interior of the convex cone M which is generated by the vectors

$$(v_1(t), \ldots, v_n(t), 1)^T, \quad (-v_1(t), \ldots, -v_n(t), 1)^T, \quad t \in T. \tag{13}$$

Note that $c \in M$, for we can pick an arbitrary $\bar{t} \in T$ and write

$$c = \tfrac{1}{2}(v_1(\bar{t}), \ldots, v_n(\bar{t}), 1)^T + \tfrac{1}{2}(-v_1(\bar{t}), \ldots, -v_n(\bar{t}), 1)^T.$$

We next assume that $c \notin \overset{\circ}{M}$ and show that a contradiction results. If $c \in M \setminus \overset{\circ}{M}$ then $c \in \mathrm{bd}\, M$ and by (3) there is a supporting hyperplane to M at c. Hence there is a vector $(y_1, \ldots, y_n, y_{n+1})^T \neq 0$ such that

$$y^T z \leq 0 = y^T c, \quad z \in M. \tag{14}$$

(We can put $\eta = 0$ since M is a cone. See (8) of §10.) Since $c = (0, \ldots, 0, 1)^T$ we find from (14) that $y_{n+1} = 0$ and hence

$$\sum_{r=1}^{n} y_r z_r \leq 0, \quad z \in M. \tag{15}$$

We observe that $(y_1, \ldots, y_n)^T \neq 0$. We have just seen that $y_{n+1} = 0$ but we know $(y_1, \ldots, y_n, y_{n+1})^T \neq 0$. We now enter the vectors (13) into (15) and arrive at

$$\sum_{r=1}^{n} y_r v_r(t) = 0, \quad t \in T,$$

contradicting the linear independence of $v_1, \ldots, v_n$ on T.
There is a simple way of imposing the condition $c \in \overset{\circ}{M}_n$.

(16) **Regularization.** Consider the problem

(P)   Minimize $\displaystyle\sum_{r=1}^{n} c_r y_r$, $a(s)^T y \geq b(s)$, $s \in S$.

Assume now that we know a solution $\bar{y}$ of (P) and a number $F > 0$ such
that $|y_r| \leq F$, $r = 1, \ldots, n$. Then we supplement the constraints of (P)
with the conditions

$$|y_r| \leq F, \quad r = 1, \ldots, n.$$

These may also be written as the (equivalent) linear constraints

$$y_r \geq -F, \quad -y_r \geq -F, \quad r = 1, \ldots, n.$$

Thus we get a modified ("regularized") problem:

(P_F)   Minimize $\displaystyle\sum_{r=1}^{n} c_r y_r$ subject to $a(s)^T y \geq b(s)$, $s \in S$,

$$\begin{aligned} e_r^T y &\geq -F \\ -e_r^T y &\geq -F \end{aligned} \quad r = 1, \ldots, n$$

where
$$e_r = (0, \ldots, 0, \underset{\uparrow}{1}, 0, \ldots, 0)^T \in R^n.$$
$$r^{th} \text{ component}$$

The vectors which define the constraints of P_F include all the unit
vectors e_r as well as all the negative unit vectors $-e_r$. Hence we find
in this case that

$$M_n = R^n$$

and the regularity condition

$$c \in \overset{\circ}{M}_n$$

is trivially met. By means of the duality theorem just proved, we find
that the dual pair (P_F) - (D_F) has no duality gap. The solvability of
(P_F) is also a consequence of this duality theorem but can alternatively
be established from the fact that the constraints of (P_F) define a compact
subset of R^n.

It is known from the Reduction Theorem (14) of §8 that every
$c \in M_n = CC(A_S)$ admits the following representation:

$$c = \sum_{i=1}^{q} a(s_i)x_i, \quad q \leq n, \tag{17}$$

where $s_1,\ldots,s_q \in S$, $x_1,\ldots,x_q > 0$ and $a(s_1),\ldots,a(s_q)$ are linearly
independent vectors. The representation (17) is generally not unique; i.e.
c can have different representations (17) and the value of q need not
be unique. A representation (17) with $q = n$ is said to be *maximal*.

(18) <u>Lemma</u>. Let c have a maximal representation; i.e.

$$c = \sum_{i=1}^{n} a(s_i)x_i, \tag{19}$$

$$x_i > 0, \quad i = 1,\ldots,n, \tag{20}$$

$$a(s_1),\ldots,a(s_n) \text{ are linearly independent.} \tag{21}$$

Then c lies in the interior $\overset{\circ}{M}_n$ of M_n.

<u>Proof</u>: Let c have the representation (19), which we write as fol-
lows:

$$c = A(s_1,\ldots,s_n)x,$$

where the matrix $A(s_1,\ldots,s_n)$ has the column vectors $a(s_1),\ldots,a(s_n)$.
$A(s_1,\ldots,s_n)$ is nonsingular by (21), so

$$x = A(s_1,\ldots,s_n)^{-1}c.$$

Let now $s_1,\ldots,s_n$ be fixed. Then the components $x_1,\ldots,x_n$ in (19) may
be looked upon as continuous functions of the vector c. From (20) we
then conclude that there is an $\varepsilon > 0$ with the property that all vectors
$\tilde{c}$ in the neighborhood $|c - \tilde{c}| \leq \varepsilon$ are such that $\tilde{x}_1,\ldots,\tilde{x}_n > 0$, where

$$\tilde{x} = A(s_1,\ldots,s_n)^{-1}\tilde{c}.$$

Thus the vector

$$\tilde{c} = A(s_1,\ldots,s_n)\tilde{x}$$

also lies in M_n. Hence there is a neighborhood of c which lies in M_n.
That is, c is in the interior $\overset{\circ}{M}_n$ of M_n, which is the desired result.

We remark that the converse of Lemma (18) is false. As an example
we consider the following 4 vectors in R^3:

$$a_1 = (0,0,1)^T,$$

$$a_2 = (1,0,1)^T,$$

$$a_3 = (1,1,1)^T,$$

$$a_4 = (0,1,1)^T.$$

Put $c = (1/2, 1/2, 1)^T$. It is easy to establish, e.g. by drawing a suitable picture, that c is in the interior of the moment cone formed by the vectors $a_1, \ldots, a_4$. Nevertheless one verifies by straightforward calculation that c has *no* representation (19) - (21) with $q = n = 3$.

We conclude this section by showing, as promised above, that

$$\overset{\circ}{M} = \overset{\circ}{M}$$

holds for nonempty convex sets $M \subset R^p$. The proof will be carried out in three steps (see also Eggleston (1958)).

(22) <u>Lemma</u>. Let $M \subset R^p$ be a nonempty set in R^p with nonempty interior $\overset{\circ}{M}$. Let x_1 and x_2 be two points in M such that $x_2 \in \overset{\circ}{M}$. Consider the line segment

$$[x_1, x_2] = \{x = \lambda x_1 + (1-\lambda)x_2 \mid \lambda \in [0,1]\}.$$

Then all of $[x_1, x_2]$, except possibly the endpoint x_1, belongs to the interior $\overset{\circ}{M}$ of M.

<u>Proof:</u> Since M is convex, $[x_1, x_2] \subset M$. $x_2 \in \overset{\circ}{M}$ implies that there is a sphere, $K_\delta(x_2)$, $\delta > 0$, with $K_\delta(x_2) \subset M$ (see (11) of §2). Let $y \neq x_1$ be a point in $[x_1, x_2]$. We want to show that there exists $r > 0$ such that

$$K_r(y) \subset M \tag{23}$$

and hence $y \in \overset{\circ}{M}$ as asserted. Put

$$y = \lambda x_1 + \mu x_2 \tag{24}$$

where $\lambda \geq 0$, $\mu \geq 0$, $\lambda + \mu = 1$. We verify now that (23) holds for $r = \mu\delta$. Let $z \in K_{\mu\delta}(y)$. Then

$$|z - y| < \mu\delta,$$

or, by (24),

$$|z - (\lambda x_1 + \mu x_2)| < \mu\delta.$$

Since $\mu > 0$ we find that

$$|(z - \lambda x_1)/\mu - x_2| < \delta;$$

i.e. $(z - \lambda x_1)/\mu$ lies in $K_\delta(x_2)$ and hence in M. Consider next the identity

$$z = \lambda x_1 + \mu(z - \lambda x_1)/\mu.$$

Due to the convexity of M, z must also belong to M, proving (23) and hence the assertion.

(25) <u>Lemma</u>. The assertion of Lemma (22) remains true when the assumption $x_1 \in M$ is replaced by the weaker requirement $x_1 \in \bar{M}$.

<u>Proof</u>: Since $x_2 \in \overset{\circ}{M}$ there is a $\delta > 0$ such that $K_\delta(x_2) \subset \overset{\circ}{M}$. Let $y \in [x_1, x_2]$ with $y \neq x_1$, $y \neq x_2$ and let z_1 be an arbitrary point in M such that

$$|z_1 - x_1| < \delta|x_1 - y|/|x_2 - y|.$$

Define z_2 through the relation

$$z_2 - x_2 = -(z_1 - x_1)|x_2 - y|/|x_1 - y|.$$

Then we obtain

$$|z_2 - x_2| < \delta, \quad \text{i.e.} \quad z_2 \in K_\delta(x_2) \subset \overset{\circ}{M}.$$

Next we find that

$$y = \lambda x_2 + \mu x_1 = \lambda z_2 + \mu z_1,$$

where

$$\lambda = |x_1 - y|/\{|x_1 - y| + |x_2 = y|\}$$

and

$$\mu = 1 - \lambda = |x_2 - \mu|/\{|x_1 - \mu| + |x_2 - \mu|\}.$$

Hence $y \in [z_2, z_1]$. Lemma (22) now delivers the desired result.

(26) <u>Theorem</u>. Let $M \subset R^p$ be a convex set with nonempty interior $\overset{\circ}{M}$. Then $\overset{\circ}{\bar{M}} = \overset{\circ}{M}$.

<u>Proof</u>: Since $M \subset \bar{M}$ we get $\overset{\circ}{M} \subset \overset{\circ}{\bar{M}}$. We establish that $\overset{\circ}{\bar{M}} \subset \overset{\circ}{M}$ by showing $x \in \bar{M}$, $x \notin \overset{\circ}{M}$ implies $x \notin \overset{\circ}{\bar{M}}$. Assume that $x \in \bar{M} \smallsetminus \overset{\circ}{M}$ and $x \in \overset{\circ}{\bar{M}}$. Select an arbitrary $x_1 \in \overset{\circ}{M}$. Since $x \in \overset{\circ}{\bar{M}}$ there is also a point $y \in \bar{M}$, $y \neq x$, with $x \in [x_1, y]$. By Lemma (25) $x \in \overset{\circ}{M}$, contradicting the assumption.

Chapter V
The Simplex Algorithm

This and the next chapter are devoted to the presentation of the simplex algorithm for the numerical solution of linear optimization problems.
This very important scheme was developed by Dantzig around 1950. We will
see that the simplex algorithm consists of a sequence of exchange steps.
A special algorithm, related to the simplex algorithm and also based on
exchange steps, was used in 1934 by Remez for the calculation of best
approximations in the uniform norm. His procedure is described in Cheney
(1966).

We will not prove the convergence of the simplex algorithm here. For
the case of finitely many constraints (linear programming) the convergence
has been established a fairly long time ago (Charnes, Cooper and
Henderson (1953), p. 62). The general case is much more difficult and
has not been studied until recently.

In this chapter we shall give a general description of the simplex
algorithm and Chapter VI will be devoted to its numerical realization.

For easy reference we state here Problem (P), which is to be treated
by means of the simplex algorithm:

$$(P) \quad \text{Minimize} \quad \sum_{r=1}^{n} c_r y_r \quad \text{subject to} \quad \sum_{r=1}^{n} a_r(s) y_r \geq b(s), \quad s \in S.$$

In this and the next chapter we shall require that (P) is solvable, if
bounded, and that no duality gap occurs. We have shown in Chapter IV, §10
that this situation occurs when M_{n+1} is closed (e.g. the case of linear
programming) or when the Slater condition is met. The dual problem can
then be written in the following form:

(D)   Maximize $\displaystyle\sum_{i=1}^{n} b(s_i)x_i$ subject to $\displaystyle\sum_{i=1}^{n} a_r(s_i)x_i = c_r,$ $r = 1,\ldots,n,$

$$s_i \in S, \quad x_i \geq 0, \quad i = 1,\ldots,n$$

(see (7), of §12). In the future we shall write a feasible solution to this problem in the form $\{\sigma,x\}$. Here, $\sigma = \{s_1,\ldots,s_n\} \subset S$ and $x = (x_1,\ldots,x_n) \in R^n$.

§12. BASIC SOLUTIONS AND THE EXCHANGE STEP

We write the constraints of (D) in the form

$$\sum_{i=1}^{n} a(s_i)x_i = c, \tag{1}$$

$$\sigma = \{s_1,\ldots,s_n\} \subset S, \quad x = (x_1,\ldots,x_n)^T \geq 0. \tag{2}$$

Here, $a(s_1),\ldots,a(s_n)$ are n of those vectors in R^n which appear in the constraints of (P):

$$a(s)^T y \geq b(s), \quad s \in S.$$

(3) <u>Definition</u>. Let $\{\sigma,x\}$ be feasible for (D), i.e. (1) and (2) hold. Also, let $a(s_1),\ldots,a(s_n)$ be linearly independent. Then $\{\sigma,x\}$ will be called a *basic solution* to (1).

Thus if $\{\sigma,x\}$ is a basic solution then the linear system of equations (1) has the unique solution x. We shall also write this system in the form

$$A(s_1,\ldots,s_n)x = c. \tag{4}$$

Here, $A(s_1,\ldots,s_n)$ is the $n \times n$ matrix having the columns $a(s_1),\ldots,$ $a(s_n)$:

$$A(s_1,\ldots,s_n) = \begin{pmatrix} a_1(s_1) & \cdots & a_1(s_n) \\ a_2(s_1) & \cdots & a_2(s_n) \\ \vdots & & \vdots \\ a_n(s_1) & \cdots & a_n(s_n) \end{pmatrix}. \tag{5}$$

Hence if $\{\sigma,x\}$ is a basic solution then the rank of this basis matrix $A(s_1,\ldots,s_n)$ is n and we have

$$x = A(s_1,\ldots,s_n)^{-1}c$$

and

$$x \geq 0.$$

(6) Requirement. We shall require that among the vectors $a(s)$,
$s \in S$, there is always a subset of n linearly independent vectors. (This
implies that if $|S| = m$, then $n \leq m$ must hold.)

(7) Lemma. Let the dual problem (D) be solvable. Then there is a
solution $\{s_1,\ldots,s_q\}$, $x_1,\ldots,x_q$ such that $q \leq n$, $x_i > 0$, $i = 1,\ldots,q$,
and the vectors

   $a(s_i)$, $i = 1,\ldots,q$

are linearly independent.

Proof: Let (D) have the value $v(D)$. Then we have the relations

$$\sum_{i=1}^{q} x_i b(s_i) = c_0 = v(D), \tag{8}$$

$$\sum_{i=1}^{q} x_i a(s_i) = c, \tag{9}$$

$$x_i \geq 0, \quad i = 1,\ldots,q.$$

Thus the vector $(c_0,\ldots,c_n) \in R^{n+1}$ is a convex combination of the q
vectors $(b(s_i),a_1(s_i),\ldots,a_n(s_i))^T \in R^{n+1}$. The representation (8), (9)
is not unique. Using the reduction theorem (14) of §8 we conclude that
among the representations (8), (9) there is at least one such that
$q \leq n+1$, $x_i > 0$, $i = 1,\ldots,q$ and $a(s_1),\ldots,a(s_q)$ are linearly independent. We now want to show that $q \leq n$. We consider therefore the moment
cone M_{n+1} which is defined as in (32) of §8. The vector $(c_0,\ldots,c_n)^T$
lies on the boundary of M_{n+1}. By Lemma (18) of §11 we must therefore
have $q \leq n$, which is the desired conclusion.

We can now state and prove an important result.

(10) Theorem (Existence of optimal basic solutions). Let the dual
problem (D) be solvable. Among the solutions there is a basic solution,
i.e. an *optimal basic* solution.

Proof: The proof is an immediate consequence of Lemma (7). There is
always a solution $\{s_1,\ldots,s_q\}$, $x_1,\ldots,x_q$ of (D) with q linearly independent vectors $a(s_1),\ldots,a(s_q)$, $q \leq n$. If $q = n$, the assertion is already established. We discuss the case $q < n$. Then we put

   $$x_{q+1} = x_{q+2} = \ldots = x_n = 0$$

and select $s_{q+1},\ldots,s_n \in S$ such that the vectors $a(s_1),\ldots,a(s_n)$ are
linearly independent. (This is always possible due to the requirement
(6).) Thus

$$\sigma = \{s_1, \ldots, s_n\} \quad \text{and} \quad x = (x_1, \ldots, x_q, 0, \ldots, 0)^T \in R^n$$

define an optimal basic solution. (This basic solution is "degenerate" in the sense of Definition (39) below.)

(11) **Definition.** The subset $\sigma = \{s_1, \ldots, s_n\} \subset S$ with exactly n elements is called a *basic set* if the matrix $A(s_1, \ldots, s_n)$ is nonsingular and the system of equations

$$A(s_1, \ldots, s_n)x = c$$

has a nonnegative solution x. (Then $\{\sigma, x\}$ is of course a basic solution of (D).)

The simplex algorithm consists of a sequence of *exchange steps*. In each step a basic set is *given* and one constructs a *new* basic set $\sigma' \subset S$ and the corresponding vector $x' \in R^n$. One seeks to achieve:

$$\sum_{i=1}^{n} b(s_i)x_i < \sum_{i=1}^{n} b(s_i')x_i'; \tag{12}$$

i.e. that $\{\sigma', x'\}$ is a better basic solution than $\{\sigma, x\}$ in the sense that the preference function of (D) assumes a larger value.

In the following we are going to split this exchange step into six substeps, each of which will be discussed in detail. Special attention will be devoted to the question of determining when an improvement (12) is possible.

The numerical considerations associated with the simplex algorithm will be dealt with in §14. In order to start the simplex algorithm an initial basic solution $\{\sigma^0, x^0\}$ must be known. In §15 we shall describe how to construct an initial basic solution.

We assume now that we are given a basic set σ and the corresponding basic solution $\{\sigma, x\}$. Thus x is the unique solution of (4).

We have already stated that the simplex algorithm also delivers approximate solutions to the primal problem (P). The following simple complementary slackness theorem indicates how the basic set σ may be associated with a vector $y \in R^n$.

(13) **Complementary slackness theorem.** Let $\bar{y}$ be feasible for (P) and $\{\bar{\sigma}, \bar{x}\}$ feasible for (D). Then $\bar{y}$ and $\{\bar{\sigma}, \bar{x}\}$ are optimal for (P) and (D) respectively if and only if

$$\bar{x}_i \left\{ \sum_{r=1}^{n} a_r(\bar{s}_i)\bar{y}_r - b(\bar{s}_i) \right\} = 0, \quad i = 1, \ldots, n. \tag{14}$$

Proof: We showed in (20) of §4 that (14) is sufficient for the optimality of $\bar{y}$ and $\{\bar{\sigma}, \bar{x}\}$. The necessity is an easy consequence of the relation

$$\sum_{i=1}^{n} b(\bar{s}_i)\bar{x}_i = v(D) = v(P) = \sum_{r=1}^{n} c_r \bar{y}_r$$

combined with the dual constraints. We recall that we have assumed $v(P) = v(D)$ in this entire chapter.

The statement of the complementary slackness theorem can also be phrased thusly: $\{\bar{\sigma}, \bar{x}\}$ and $\bar{y}$ are optimal for the Problems (P) and (D) respectively if and only if they satisfy the following systems of equations and inequalities:

Primal constraints

$$\sum_{r=1}^{n} a_r(s)y_r \geq b(s), \quad s \in S. \tag{15}$$

Dual constraints

$$\sum_{i=1}^{n} a_r(s_i)x_i = c_r, \quad r = 1,\ldots,n \tag{16}$$

$$x_i \geq 0, \quad i = 1,\ldots,n.$$

Complementary slackness conditions

$$x_i \left\{ \sum_{r=1}^{n} a_r(s_i)y_r - b(s_i) \right\} = 0, \quad i = 1,\ldots,n. \tag{17}$$

Our given basic solution $\{\sigma, x\}$ must of course satisfy (16).

Starting from $\{\sigma, x\}$ we determine a vector $y \in R^n$ such that (15) is satisfied as well by selecting y as the solution of the equations

$$\sum_{r=1}^{n} a_r(s_i)y_r = b(s_i), \quad i = 1,\ldots,n.$$

This system has a unique solution y since the system can be written

$$A^T(s_1,\ldots,s_n)y = b(s_1,\ldots,s_n). \tag{18}$$

Here $A^T(s_1,\ldots,s_n)$ is the transpose of the matrix $A(s_1,\ldots,s_n)$ in (5) and

$$b(s_1,\ldots,s_n) = (b(s_1),\ldots,b(s_n))^T \in R^n.$$

$A(s_1,\ldots,s_n)$ is nonsingular by the definition of basic solution. Hence $A^T(s_1,\ldots,s_n)$ has the same property. Thus

$$y = A^T(s_1,\ldots,s_n)^{-1}b(s_1,\ldots,s_n)$$

is uniquely determined by (18).

    (19) <u>Exchange Substeps (E1) and (E2)</u>. The basic set

$$\sigma = \{s_1,\ldots,s_n\} \subset S$$

is given.

    (E1) Compute the unique nonnegative solution x to the linear system of equations

$$A(s_1,\ldots,s_n)x = c.$$

    (E2) Determine the unique solution y to the linear system of equations

$$A^T(s_1,\ldots,s_n)y = b(s_1,\ldots,s_n).$$

If y also satisfies

$$\sum_{r=1}^{n} a_r(s)y_r \geq b(s), \quad s \in S,$$

then y is optimal for (P) and $\{\sigma,x\}$ optimal for (D). We assume now that we are given a basic set

$$\sigma = \{s_1,\ldots,s_n\} \subset S$$

such that the vector y calculated in (E2) does not meet all the conditions (15). Then

$$\{\sigma,x,y\}$$

is *not* a solution to the system (15) - (17). We describe now how to find an approximate solution

$$\{\sigma',x',y'\}$$

to the system of equations and inequalities (15) - (17) which is better in the sense of (12). The basic sets σ and σ' will have all elements except one in common. Thus if

$$\sigma = \{s_1,\ldots,s_n\},$$

then exactly one s_i, $i = 1,\ldots,n$, say s_r, will be exchanged for an $s' \in S$ which did not belong to σ. Hence

$$\sigma' = \{s_1,\ldots,s_{r-1},s', s_{r+1},\ldots,s_n\} = \{s_1',\ldots,s_n'\}$$

where

$$s_i' = s_i, \quad , \; i \neq r$$

$$s_r' = s'.$$

Alternatively,

$$\sigma' = (\sigma \cup \{s'\}) \smallsetminus \{s_r\}$$

for some $r \in \{1,\ldots,n\}$. We describe first how to select $s' \in S$ to be *included* in σ'. $\{\sigma,x,y\}$ are hence given.

(20)   Exchange Substep (E3).

(E3)   Determine $s' \in S$ such that

$$\sum_{r=1}^{n} a_r(s')y_r < b(s'). \tag{21}$$

If no such s' exists, then the computation is stopped here, since $\{\sigma,x,y\}$ solves (15) - (17).

This means that we include in the basic set σ' a point s' which is such that a primal constraint is violated. This fact entails that $s' \notin \sigma$.

There remains to determine a member $s_i \in \sigma$ which shall leave σ, i.e. will be replaced by s'.

(22)   Exchange Substep (E4).

(E4)   Compute the solution $d \in R^n$ of the linear system of equations

$$A(s_1,\ldots,s_n)d = a(s') \tag{23}$$

i.e.

$$\sum_{i=1}^{n} a(s_i)d_i = a(s').$$

(23) thus expresses the "new" vector $a(s')$ as a linear combination of the "old" vectors $a(s_i)$, $s_i \in \sigma$. The meaning of the vector d is clear from the following argument. Consider the set

$$\sigma \cup \{s'\} = \{s_1,\ldots,s_n,s'\} \subset S. \tag{24}$$

It consists of $n+1$ elements. Introduce the n+1-dimensional vector

$$(x_1 - \lambda d_1,\ldots,x_n - \lambda d_n,\lambda)$$
$$= (x_1(\lambda),\ldots,x_n(\lambda), x_{n+1}(\lambda))^T \in R^{n+1} \tag{25}$$

($\lambda \in R$ is arbitrary). The value of the dual preference function for $\{\sigma \cup \{s'\}, x(\lambda)\}$ will be denoted by $c_0(\lambda)$:

$$c_0(\lambda) = \sum_{i=1}^{n} b(s_i)x_i(\lambda) + b(s')\lambda = \sum_{i=1}^{n} b(s_i)(x_i - \lambda d_i) + b(s')\lambda.$$

If we put $\lambda = 0$, we get

$$c_0(0) = \sum_{i=1}^{n} b(s_i)x_i,$$

the "old" value of the dual preference function.

(26) <u>Lemma</u>. The following relation is true for all λ:

$$c_0(\lambda) = c_0(0) + \lambda\Delta(s'), \tag{27}$$

where

$$\Delta(s') = b(s') - \sum_{r=1}^{n} a_r(s')y_r > 0.$$

(Compare (21).)

<u>Proof</u>: Using (18) and (23) we have

$$
\begin{aligned}
c_0(\lambda) &= \sum_{i=1}^{n} b(s_i)x_i + \lambda\{b(s') - \sum_{i=1}^{n} b(s_i)d_i\} \\
&= c_0(0) + \lambda\{b(s') - b(s_1,\ldots,s_n)^T d\} \\
&= c_0(0) + \lambda\{b(s') - y^T A(s_1,\ldots,s_n)d\} \\
&= c_0(0) + \lambda\{b(s') - y^T a(s')\} \\
&= c_0(0) + \Delta\lambda(s').
\end{aligned}
$$

Since $\Delta(s') > 0$, the value of the dual preference function for $x(\lambda)$ is not smaller than that for $x = x(0)$. Therefore, if $\{\sigma \cup \{s'\}, x(\lambda)\}$ is feasible for *all* $\lambda \geq 0$, then the value of the dual preference function can be made arbitrarily large. This should mean that (D) is unbounded, entailing that (P) is inconsistent. This case is dealt with in the following lemma.

(28) <u>Lemma</u>. Let the unique solution vector d of (23) be such that

$$d_i \leq 0, \quad i = 1,\ldots,n. \tag{29}$$

Then (D) is unbounded and hence (P) is inconsistent.

<u>Proof</u>: We note first that (23), (24) and (25) imply that the equality constraints of the dual problem are met independently of (29). Thus

$$\sum_{i=1}^{n} a_r(s_i)x_i(\lambda) + a_r(s')x_{n+1}(\lambda) = c_r, \quad r = 1,\ldots,n,$$

and this equation is true for all real λ. If (29) holds as well, then

$$x_i(\lambda) \geq 0, \quad i = 1,\ldots,n,$$

for all $\lambda \geq 0$. Letting $\lambda \to +\infty$, by (27) we conclude that $c_0(\lambda) \to +\infty$, establishing the assertion.

It is now clear how to select λ when some of the d_i are positive. One calculates the *maximal* λ such that

$$x_i(\lambda) = x_i - \lambda d_i \geq 0, \quad i = 1,\ldots,n. \tag{30}$$

Then one need only consider those indices i such that $d_i > 0$. If $d_i > 0$, then (30) is equivalent to

$$\lambda \leq x_i/d_i.$$

Thus

$$\bar{\lambda} = \min \{x_i/d_i, \ d_i > 0\}$$

meets all the conditions (30). It is also clear that *at least one* of the components $x_i(\bar{\lambda})$, $i = 1,\ldots,n$ of the vector $x(\bar{\lambda})$ will vanish. Indeed, if

$$\bar{\lambda} = x_r/d_r \quad \text{for an} \quad r \in \{1,\ldots,n\}, \tag{31}$$

then we get

$$x_r(\bar{\lambda}) = x_r - \frac{x_r}{d_r} d_r = 0. \tag{32}$$

The corresponding element s_r *is removed from the basic set.* Hence we put

$$\sigma' = \{\sigma \cup \{s'\}\} \smallsetminus \{s_r\}$$

$$= \{s_1,\ldots,s_{r-1},s', s_{r+1},\ldots,s_n\}$$

$$= \{s_1',\ldots,s_{r-1}',s_r', s_{r+1}',\ldots,s_n'\}$$

and

$$x' = (x_1(\bar{\lambda}),\ldots,x_{r-1}(\bar{\lambda}),\bar{\lambda}, x_{r+1}(\bar{\lambda}),\ldots,x_n(\bar{\lambda}))^T$$

$$= \left(x_1 - \frac{x_r}{d_r} d_1,\ldots,x_{r-1} - \frac{x_r}{d_r} d_{r-1}, \frac{x_r}{d_r}, x_{r+1} - \frac{x_r}{d_r} d_{r+1},\ldots,\right.$$

$$\left.\ldots x_n - \frac{x_r}{d_r} d_n\right)^T. \tag{33}$$

(34) **Exercise.** Use (27) and (32) to verify once more that $\{\sigma',x'\}$ is feasible for (D) and that

$$\sum_{i=1}^{n} b(s_i')x_i' = \sum_{i=1}^{n} b(s_i)x_i + \frac{x_r}{d_r} \Delta(s').$$

(31) does not necessarily determine the index $r \in \{1,\ldots,n\}$ uniquely.

We summarize the process above (i.e. the determination of which element s_r to remove from σ) as follows:

(35) **Exchange Substeps (E5) and (E6).** Let d be the unique solution of (23) in Substep (E4).

  (E5) If $d_i \leq 0$, $i = 1,\ldots,n$, then (D) is unbounded and (P) is inconsistent. The computations are stopped.

  (E6) If there is a positive d_i, then select an $r \in \{1,\ldots,n\}$ with $d_r > 0$ and such that

$$\frac{x_r}{d_r} = \min\{x_i/d_i, \ d_i > 0\}.$$

Next put

$$\sigma' = \{\sigma \cup \{s'\}\} \smallsetminus \{s_r\}.$$

Now the fundamental question arises whether the "new" set σ' is a basic set. In that case one can repeat the process from Substep (E1) (with σ' instead of σ). Thus one gets an iterative scheme, the simplex algorithm. We now prove

(36) **Lemma.** Let s' be found via Substep (E3) and s_r via Substep (E6). Then σ' is a basic set.

Proof: To facilitate the presentation we renumber the vectors $a(s_i)$, $i = 1,\ldots,n$ so that $r = 1$. Thus we must show that

$$a(s'), \ a(s_2),\ldots,a(s_n) \tag{37}$$

are linearly independent. Since σ is a basic set the vectors

$$a(s_2),\ldots,a(s_n) \tag{38}$$

must be linearly independent. Assume that $a(s')$, $a(s_2),\ldots,a(s_n)$ are linearly dependent. Then $a(s')$ can be written as a linear combination of the vectors in (38);

$$a(s') = \sum_{i=2}^{n} a(s_i)\rho_i.$$

Comparing with (23) we find that

$$d_1 = 0, \quad d_2 = \rho_2, \ldots, d_n = \rho_n.$$

This contradicts the fact that we have assumed $r = 1$ since r is always selected such that $d_r > 0$. The system of equations

$$A(s_1, \ldots, s_{r-1}, s', s_{r+1}, \ldots, s_n)x' = c$$

has a unique *nonnegative* solution x' since the index r was selected in Substep (E6) precisely according to that criterion. (See also (33).)

Thus Lemma (36) guarantees that one can return to Substep E1 with the new basic set σ', provided no interruption occurs in Substeps (E3) or (E5). As stated earlier, the goal is to increase the dual preference function, i.e. to achieve that

$$b(s_1, \ldots, s_n)^T x < b(s_1', \ldots, s_n')x'$$

holds at each simplex step. Unfortunately this *cannot* be provided for under all circumstances. That is, if

$$b(s_1', \ldots, s_n')^T x' = b(s_1, \ldots, s_n)^T x + \frac{x_r}{d_r} \Delta(s').$$

and s' and s_r are chosen such that

$$\Delta(s') > 0 \quad \text{and} \quad d_r > 0$$

then it is quite possible that

$$x_r = 0$$

holds. Then the value of the dual preference function would remain constant during the transfer from the basic set σ to the new basic set σ'. Such an exchange would appear not to be worthwhile.

(39) <u>Definition</u>. A basic solution $\{\sigma, x\}$ is termed *regular* if $x_i > 0$, $i = 1, \ldots, n$. If at least one $x_i = 0$, then the basic solution is called *degenerate*.

(40) <u>Exercise</u>. We are given the following optimization problem (P)

(P)   Minimize $\displaystyle\sum_{r=1}^{6} \frac{1 + (-1)^{r-1}}{r} y_r$ subject to

$$\sum_{r=1}^{6} s^{r-1} y_r \geq e^s, \quad s \in [-1, 1].$$

The corresponding dual problem reads

$$\text{Maximize} \quad \sum_{i=1}^{q} e^{s_i} x_i \quad \text{subject to}$$

(D)
$$\sum_{i=1}^{q} s_i^{r-1} x_i = \frac{1 + (-1)^{r-1}}{r}, \quad r = 1,\ldots,6,$$

$$x_i \geq 0, \quad i = 1,\ldots,q.$$

Verify the statements below.

   i) Put $q = 7$ and define $\sigma^{(1)} = \{s_1,\ldots,s_7\}$, $x^{(1)} \in \mathbb{R}^7$ by

$$\sigma^{(1)} = \{-1, -\sqrt{3/5}, -\sqrt{1/5}, 0, \sqrt{1/5}, \sqrt{3/5}, 1\},$$

$$x^{(1)} = (1/12, 5/18, 5/12, 4/9, 5/12, 5/18, 1/12)^T.$$

Then $\{\sigma^{(1)}, x^{(1)}\}$ is feasible for (D) but is not a basic solution.

   ii) Let

$$\sigma^{(2)} = \{-1, -\sqrt{3/5}, -\sqrt{1/5}, \sqrt{1/5}, \sqrt{3/5}, 1\},$$

$$x^{(2)} = (\frac{19}{144}, \frac{25}{48}, \frac{25}{72}, \frac{25}{72}, \frac{25}{48}, \frac{19}{144})^T.$$

Then $\{\sigma^{(2)}, x^{(2)}\}$ is a regular basic solution.

   iii) Using the reduction process from (14) of §8, one may construct
   from $\{\sigma^{(1)}, x^{(1)}\}$ a basic solution with the basic set

$$\sigma^{(3)} = \{-1, -\sqrt{3/5}, 0, \sqrt{1/5}, \sqrt{3/5}, 1\},$$

$$x^{(3)} = (0, 5/9, 8/9, 0, 5/9, 0)^T.$$

Then $\{\sigma^{(3)}, x^{(3)}\}$ is a degenerate basic solution.

We observe that when an optimization problem is such that all basic
sets are regular then the dual preference function increases with each
simplex step.

We now summarize all the Substeps of the exchange step for the linear
optimization problems of type (P).

   (41) <u>The exchange step of the simplex algorithm.</u> Let a basic set

$$\sigma = \{s_1,\ldots,s_n\} \subset S$$

be given (the construction of an initial basic set is treated in §15). We
introduce the nonsingular matrix

$$A(s_1,\ldots,s_n)$$

with the columns $a(s_1),\ldots,a(s_n)$, and the vector

$$b(s_1,\ldots,s_n) = (b(s_1),\ldots,b(s_n))^T.$$

(E1) Determine $x \in R^n$ from

$$A(s_1,\ldots,s_n)x = c.$$

(E2) Compute $y \in R^n$ from

$$A^T(s_1,\ldots,s_n)y = b(s_1,\ldots,s_n).$$

(E3) Determine an $s' \in S$ such that

$$\sum_{r=1}^{n} a_r(s')y_r < b(s').$$

If no s' with this property exists, then y is optimal for (P) and
$\{\sigma,x\}$ optimal for (D), and the calculations are stopped here.

(E4) Compute $d = (d_1,\ldots,d_n)^T \in R^n$ such that

$$A(s_1,\ldots,s_n)d = a(s').$$

(E5) If

$$d_i \leq 0, \quad i = 1,\ldots,n,$$

then (D) is unbounded and (P) is inconsistent, and the computations are stopped here.

(E6) Find $r \in \{1,\ldots,n\}$ such that

$$\frac{x_r}{d_r} = \min \left\{ \frac{x_i}{d_i} \Big/ d_i > 0 \right\}$$

and put $\sigma' = \{\sigma \cup \{s'\}\} \smallsetminus \{s_r\}$, i.e.

$$\sigma' = \{s_1,\ldots,s_{r-1},s', s_{r+1},\ldots,s_n\} = \{s'_1,\ldots,s'_n\}.$$

Then σ' is a basic set and the corresponding basic solution
satisfies

$$b(s'_1,\ldots,s'_n)^T x' = b(s_1,\ldots,s_n)^T x + \frac{x_r}{d_r} \Delta(s').$$

(42) Remark. The Substeps (E1), (E2) and (E4) call for the solution
of linear systems of equations. We have not yet described how to arrange
the calculations efficiently. The different *variants* of the simplex algorithm differ only in this respect. Fundamental for the analysis of the
numerical properties of the various simplex algorithms is the recognition

that at each simplex iteration linear systems of equations are solved, explicitly or implicitly. We shall discuss this matter in §14.

(43) <u>Remark</u>. We note that exactly *one* element is exchanged by the transfer from the "old" basic set σ to the "new" one σ'. There are other exchange procedures by which *several* elements are exchanged at each step. One extreme case is the so-called simultaneous exchange when all elements of σ are changed by the transfer to σ' (see Judin and Goldstein (1968), p. 506). We also mention in this context the Remez algorithm (see Cheney (1966), p. 97), where again the entire basic set is exchanged at each step. The computational effort is generally greater than by the exchange algorithm described above but on the other hand one hopes to achieve greater increases in the value of the dual preference function per iteration step.

§13. THE SIMPLEX ALGORITHM AND DISCRETIZATION

Let an initial basic set $\sigma^0 = \{s_1^0, \ldots, s_n^0\}$ be known. (See §15.) If we now perform an exchange step and no interruption occurs in (E3) and (E5) (in each of these cases there is no need to continue the computations), then (E6) gives a new basic set $\sigma' = \{s_1^1, \ldots, s_n^1\}$. Hence we can return to Substep (E1) and start a new exchange step. In this way we have obtained the *simplex algorithm*. Thus we generate a sequence

$$\sigma^0 \to \sigma^1 \to \sigma^2 \ \ldots$$

of basic sets,

$$\sigma^k = \{s_1^k, \ldots, s_n^k\}, \quad k = 0, 1, \ldots \ .$$

Note that σ^k and σ^{k+1} have all elements except exactly one in common. We also get a corresponding sequence of basic matrices

$$A_0 \to A_1 \to A_2 \to \ldots \ ,$$

where

$$A_k = A(s_1^k, \ldots, s_n^k)$$

has the column vectors $a(s_1^k), \ldots, a(s_n^k)$. The corresponding vectors

$$x^k = A_k^{-1} c, \quad k = 1, 2, \ldots$$

are such that

$$b_1^T x^1 \leq b_2^T x^2 \leq \cdots \leq b_k^T x^k \leq b_{k+1}^T x^{k+1} \leq \cdots \leq v(D),$$

where

$$b_k = b(s_1^k, \ldots, s_n^k).$$

(1) <u>Remark</u>. The matrix A_{k-1} differs from A_k only by one column vector!

We now want to describe in greater detail how to determine the vector $a(s')$ which is to be included in the basis (Substep (E3)). There are in general very many indices $\bar{s} \in S$ such that

$$\sum_{r=1}^{n} a_r(\bar{s}) y_r - b(\bar{s}) < 0.$$

If one wants to write a computer program for carrying out the exchange step, then one must given an unambiguous selection rule.

(2) <u>The case of linear programming, $|S| < \infty$</u>. In this case S is a finite set. Usually one has the rule to select s' at the minimum point of the error function

$$\sum_{r=1}^{n} a_r(s) y_r - b(s).$$

Thus we take an index value which renders the function

$$\Delta(s) = b(s) - \sum_{r=1}^{n} a_r(s) y_r$$

a maximum. Hence, in Exchange Substep (E3) we add to the basis an element of S which is such that the primal constraints are violated as much as possible. Since S is finite we can determine an element s' which has the property

$$\Delta(s') \geq \Delta(s), \quad s \in S \tag{3}$$

by means of finitely many arithmetic operations. If s' is not uniquely defined by (3), then we must introduce further conventions to make a unique choice possible. If S is an ordered set, e.g. a finite subset of a real interval, we take as s' the smallest index satisfying (3).

Thus the Substep (E3) of the Exchange step is completely specified for a *finite* index set. For this class of linear optimization problems one can establish a simple result on the convergence of the simplex algorithm.

Consider the case $S = \{1,\ldots,m\}$ where $m \geq n$. Then there are only
finitely many different basis sets $\sigma = \{s_1,\ldots,s_n\}$. Indeed, there are

$$\gamma_{m,n} = \binom{m}{n} = \frac{m!}{n!\,(m-n)!}$$

different subsets of S with n elements. Hence there are *at most* $\gamma_{m,n}$
different basic solutions of the system occurring in the dual problem (LD)

$$Ax = c, \quad x \geq 0.$$

In principle, it is possible to solve the dual pair (LP) - (LD) by
means of calculating all these basic solutions and then to pick the one
which assigns the highest value to dual preference function.

In practice this is not possible since the computational effort
thereby required is prohibitive even for modest values of m and n. As
an example we mention that

$$\gamma_{20,10} = 184756.$$

The decisive advantage of the simplex algorithm is the fact that a sequence
of basic solutions is *systematically* generated in such a manner that the
corresponding values of the dual preference function form a nondecreasing
sequence. Therefore usually only a small fraction of the possible number
of basic sets will be generated. This is the reason for the efficiency
of the simplex algorithm of linear programming.

(4) **Theorem.** Let S have finitely many elements; i.e. we consider
the dual pair (LP) - (LD) of linear programs. Let (LD) be feasible and
bounded. Assume also that the simplex algorithm generates a sequence of
basic solutions such that the corresponding values of the dual preference
function form a strictly increasing sequence. Then the simplex algorithm
delivers optimal solutions to (LP) and (LD) after finitely many iterations.

Proof: Since the values of the preference function corresponding to
the basic solutions which are generated by the simplex algorithm are
strictly increasing, the same basic set cannot appear twice. Thus the
simplex algorithm generates pairwise different basic sets. Since there
are only finitely many basic sets the simplex algorithm must stop at an
optimal solution after finitely many iterations.

(5) **Remark.** If all the basic solutions which are generated by the
simplex algorithm are regular (see (39) of §12), then the preference function of the dual problem is strictly increasing. Hence the simplex algorithm must deliver an optimal solution after finitely many iterations.

If degenerate basic solutions occur, it is quite possible that the
simplex algorithm "cycles", i.e. the same basic solutions reappear
periodically and the value of the dual preference function remains constant
without having reached its optimum. Examples illustrating this phenomena
have been constructed. However, such "pathological" cases occur so rarely
that one generally does not bother with taking special precautions for
dealing with them when constructing computer programs for practical use.
It sometimes happens that degenerate basic sets do occur and thus one or
several simplex steps are carried out through which the current value of
the dual preference function does not increase, but normally the increase
resumes without the use of any special devices for achieving this desired
state of affairs.

The case of degeneracy and possible cycling is of course of great
theoretical interest. By means of a modification of Exchange Substep (E6)
the simplex algorithm may be altered so that the same basic set cannot
reappear even if degeneracy occurs. Then the simplex algorithm gives an
optimal solution after finitely many iterations in this more general situation as well. The principle behind this modification is to introduce an
arbitrary small perturbation of the vector c in the primal preference
function. Hence we construct a perturbed problem such that no degenerate
basic solutions are generated by the simplex method. Hence this perturbed
problem is solved after finitely many simplex iterations. By construction
one can now determine an optimal solution of the original problem from the
calculated optimal solution of the perturbed problem. This so-called
ε-method is described in Charnes, Cooper and Henderson (1953). It uses
the so-called lexicographic ordering to modify Exchange Substep (E6).
See also Hadley (1964) or Collatz and Wetterling (1971).

It is much more difficult to prove a convergence statement of the
form

$$\lim_{k \to \infty} b_k^T x^k = v(D)$$

when there are infinitely many constraints. Then the simplex algorithm
can not, in general, be shown to stop after finitely many iterations.
Theoretical investigations of this case can be found in the book by Blum
and Oettli (1975), p. 247-255 and in the writings by Carasso (1973) and
Hofmann and Klostermair (1976).

When S has infinitely many elements, then there is of course *no*
general procedure to find an s' satisfying (3). Without special assumptions on the index set S and the functions a_r, r = 1,...,n and b, it

is not certain that an s' with the property (3) *exists*. Even for the
special case when S is a compact subset of R^k and $a_1,\ldots,a_n,b$ are
continuous on S, it is not possible to give a general method to determine an s where $\Delta(s)$ assumes its maximum value. The case just mentioned has appeared several times before in our text. It often occurs
in uniform approximation problems.   In theoretical analysis (e.g. convergence proofs) one often works with s' satisfying (3). Some minor
relaxations of this condition are sometimes introduced. But in practice
one normally proceeds along the lines to be given below:

(6) <u>Modification of the exchange substep (E3) when</u> $|S| = \infty$. Select
a finite subset $\{s_1,\ldots,s_m\} \subset S$ and determine an s' such that

$$\Delta(s') \geq \Delta(s), \quad s \in S_m. \tag{7}$$

(If s' is not uniquely determined by (7), then one proceeds as described
in (2).)

It is easy to realize that this corresponds to a *discretization* of
(P) in the sense of (10) of §3. Consider the linear program

(P_m) Minimize $c^T y$ subject to $a(s)^T y \geq b(s)$, $s \in S_m$.

If we now start with a basis $\sigma \in S_m$ and use the selection rule from (2)
then the simplex algorithm applied to (P_m) above delivers the same new
basis elements s' as when it is used on the continuous problem

(P)   Minimize $c^T y$ subject to $a(s)^T y \geq b(s)$, $s \in S$

when one also starts from σ and determines s' according to (7).
The "rough" calculation of the new element s' to enter the basis
and (approximately) satisfying

$$\Delta(s') \geq \Delta(s), \quad s \in S$$

thus corresponds to a discretization of (P). This gives us a reason to
discuss discretization of linear optimization problems with infinitely
many constraints. Discretization is very important, both in theory and in
practice.

Consider the problem

(P)   Minimize $c^T y$ subject to $\sum_{r=1}^{n} a_r(s) y_r \geq b(s)$, $s \in S$.

This problem is approximated by the linear program

(P_m) Minimize $c^T y$ subject to $\displaystyle\sum_{r=1}^{n} a_r(s_i)y_r \geq b(s_i)$, $i = 1,\ldots,m$.

Here, $\{s_1,\ldots,s_m\}$ is a fixed subset of S.

We now give a useful interpretation of the discretized program (P_m). S is assumed to be a subset of R^k.

(8) Definition. Let $T = \{s_1,\ldots,s_m\}$ be a subset of S, and $w_1,\ldots,w_m$ be real-valued functions with the properties (i) and ii) below:

i) $w_j(s) \geq 0$, $s \in S$, $j = 1,\ldots,m$;

ii) $w_j(s_i) = \begin{cases} 1, & i = j, \\ 0, & i \neq j, \end{cases}$ $i,j = 1,\ldots,m$.

Suppose a real-valued function f is defined on S. We define the new function $Lf: S \rightarrow R$ by

$$(Lf)(s) = \sum_{j=1}^{m} w_j(s)f(s_j).$$

Then L is called a *positive interpolating operator* with *nodes* $s_1,\ldots,s_m$.

(9) Example. Piecewise linear interpolation in one dimension; $S = [\alpha,\beta]$, $\alpha = s_1 < s_2 < \ldots < s_m = \beta$. Define w_j, $j = 1,\ldots,m$ according to:

$$w_j(s) = \begin{cases} 0 & \alpha \leq s \leq s_{j-1} \quad \text{(only for } j > 1) \\ (s-s_{j-1})/(s_j-s_{j-1}) & s_{j-1} \leq s \leq s_j \\ (s_{j+1}-s)/(s_{j+1}-s_j) & s_j \leq s \leq s_{j+1} \\ 0 & s_{j+1} \leq s \leq \beta \quad \text{(only for } j < m) \end{cases}$$

See Fig. 13.1. This construction may be generalized to the "triangulation" of multidimensional areas. It is also possible to work with weighting functions w_j of a more general nature, e.g. piecewise polynomials of degree higher than 1.

The following result motivates the use of positive interpolating operators.

(10) Theorem. Let L be a positive interpolating operator with nodes $s_1,\ldots,s_m$. Then the linear optimization problem

(P_L) Minimize $c^T y$ subject to $\displaystyle\sum_{r=1}^{n} (La_r)(s)y_r \geq (Lb)(s)$, $s \in S$

has the same feasible vectors y and hence the same solution as the dis-

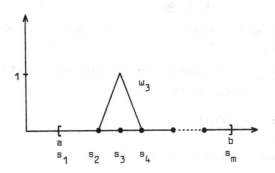

Fig. 13.1

cretized problem (P_m).

    <u>Proof</u>: a) Let y meet the constraints of (P_L). Since

$$(Lf)(s_i) = f(s_i), \quad i = 1,\ldots,m,$$

we find that y also satisfies the constraints of (P_m).

    b) Assume on the other hand that

$$\sum_{r=1}^{n} a_r(s_i)y_r \geq b(s_i), \quad i = 1,\ldots,m.$$

Since $w_i(s) \geq 0$, $i = 1,\ldots,m$ and $s \in S$, we get

$$\sum_{r=1}^{n} (La_r)(s)y_r - (Lb)(s) = \sum_{i=1}^{m} w_i(s)\left\{ \sum_{r=1}^{n} y_r a_r(s_i) - b(s_i)\right\} \geq 0$$

for all $s \in S$, proving the assertion.

    The discretization (P_m) of (P) is equivalent to replacing (P) by a
linear optimization problem with the same index set S but with the functions a_r, b approximated by La_r, Lb respectively. It is possible to
express the deviation of the optimal value of (P_m) from that of (P) in
terms of the interpolation errors

$$\max_{s\in S}|La_r(s) - a_r(s)|, \quad r = 1,\ldots,n, \quad \max_{s\in S}|Lb(s) - b(s)|.$$

Compare Theorem (16)!

    (11) <u>Definition</u>. Let S be a subset of R^k and let $\{s_1,\ldots,s_m\} \subset S$
be a grid. Let

$$h = h(s_1,\ldots,s_m) = \max_{s \in S} \min_{1 \le i \le m} |s - s_i|.$$

Then h is called the *roughness* of the grid. Here, $|\ \ |$ denotes the Euclidean distance in R^k.

(12) **Exercise.** Consider the interpolating operator of (9). Show that there is a constant c such that

$$\max_{s \in [\alpha,\beta]} |f(s) - (Lf)(s)| \le ch^2$$

when f is twice continuously differentiable.

Note. This result *cannot* be directly generalized to R^k, k > 1.

(13) **Two numerical examples.**

a) Minimize $y_1 + y_2/2 + y_3/2 + y_4/3 + y_5/4 + y_6/3$ subject to the constraints

$$y_1 + y_2 s + y_3 t + y_4 s^2 + y_5 st + y_6 t^2 \ge e^{s^2 + t^2}, \quad s \in [0,1], \quad t \in [0,1]$$

$$|y_r| \le 10, \quad r = 1,\ldots,6.$$

The index set $[0,1] \times [0,1]$ is replaced by the 25 points (s_i, t_j) where $s_i = 0.25\,(i-1)$, $t_j = 0.25\,(j-1)$, $i = 1,\ldots,5$, $j = 1,\ldots,5$. The roughness of the grid is $h = 0.125\sqrt{2} \approx 0.1768$. The discretized problem is hence a linear program with 6 variables and 37 constraints. It was solved by means of the simplex method. In the table below the solutions of the discretized and the continuous problems are given. The latter was solved with the three-phase algorithm of Chapter VII.

	Discretized Problem	Original Problem
Optimal value	2.41	2.44
Optimal solution		
y_1	2.86	2.58
y_2	-4.69	-4.11
y_3	-4.69	-4.11
y_4	4.55	4.25
y_5	4.31	4.53
y_6	4.55	4.25

This example was solved by means of the computer codes of K. Fahlander (1973).

b) The following example gives an idea about how rapidly the discretization error decreases when the grid is refined. We consider the
problem

$$\text{Minimize } \sum_{r=1}^{6} y_r/r \text{ subject to } \sum_{r=1}^{6} s^{r-1} y_r \geq 1/(1+s^2), \quad 0 \leq s \leq 1.$$

We discretize this problem by replacing the index set S by the subset

$$S_\ell = \{s_{i\ell} = (i-1)/(\ell-1), \quad i = 1,\ldots,\ell\}.$$

Using the simplex algorithm we got the results below. (The original
problem was again solved by means of the three-phase algorithm of Chapter
VII.)

Index Set	Roughness of Grid	Optimal Value
S_{21}	1/40	0.785 561 34
S_{41}	1/80	0.785 568 72
S_{81}	1/160	0.785 568 92
S	-	0.785 569 11

(14) <u>Solution of linear optimization problems by discretizations.</u>
Select a *sequence* of finite subsets S_ℓ, $\ell = 1,\ldots$ of the index set S
with the properties

$$h(S_\ell) = \max_{s \in S} \min_{\tilde{s} \in S_\ell} |s - \tilde{s}| \to 0 \quad \text{when } \ell \to \infty, \tag{15}$$

and

$$S_\ell \subset S_{\ell+1}, \quad \ell = 1,2,\ldots .$$

The linear programs (P_ℓ) are solved by means of the simplex algorithm:

$$(P_\ell) \quad \text{Minimize } c^T y \text{ subject to } \sum_{r=1}^{n} a_r(s) y_r \geq b(s), \quad s \in S .$$

An optimal basic solution to the dual (D_ℓ) can be used as the starting
basic solution to $D_{\ell+1}$.

Remark. It is possible to prove that

$$\lim_{\ell \to \infty} v(P_\ell) = v(P)$$

provided that the assumptions of the duality theorem (7) of §11 are met,
the sequence of discretizations satisfies (15), $S \subset R^k$ is a compact set,
and the functions $a_1,\ldots,a_n,b$ are continuous on S.

The following simple theorem can often be used to estimate the difference between the optimal value of the discretized problem and that of the original problem.

(16) <u>Theorem</u>. Let the linear optimization problem be such that there is a vector $\hat{y} \in R^n$ and a real number $\rho > 0$ satisfying

$$a(s)^T\hat{y} = \sum_{r=1}^{n} a_r(s)\hat{y}_r \geq \rho, \quad s \in S.$$

Let $\{s_1, \ldots, s_m\}$ be a subset of S. The linear program arising when S is replaced by this subset is assumed to have a solution $y^{(m)}$. Let $\Delta_m > 0$ be such that

$$\sum_{r=1}^{n} a_r(s)y_r^{(m)} + \Delta_m \geq b(s), \quad s \in S. \tag{18}$$

Then $v(P)$, the value of the linear optimization problem (P), can be bracketed as follows:

$$c^T y^{(m)} \leq v(P) \leq c^T y^{(m)} + \Delta_m \rho^{-1} c^T \hat{y}$$

<u>Proof</u>: The leftmost inequality is well known. See (12) of §3. To show the other inequality we observe that the vector

$$y = y^{(m)} + \Delta_m \rho^{-1}\hat{y}$$

meets the conditions of (P). We find from (17) and (18) that

$$\sum_{r=1}^{n} a_r(s)y_r = \sum_{r=1}^{n} a_r(s)y_r^{(m)} + \Delta_m \rho^{-1} \sum_{r=1}^{n} a_r(s)\hat{y}_r \geq b(s), \quad s \in S.$$

Hence we get

$$v(P) \leq c^T y = c^T y^{(m)} + \Delta_m \rho^{-1} c^T \hat{y},$$

establishing the desired result.

Chapter VI

Numerical Realization of the Simplex Algorithm

In this chapter we shall describe how to implement the simplex algorithm on a computer. As stated earlier, this algorithm requires the solution of a sequence of linear systems of equations. We devote considerable space to explaining how to solve such systems in a computationally efficient way. In the last section we discuss the construction of a basic solution with which one can start the simplex algorithm.

§14. STABLE VARIANTS OF THE SIMPLEX ALGORITHM

Each exchange step of the simplex algorithm calls for the solution of three linear systems of equations. In Substeps (E1), (E2) and (E4) we encounter

$$A_k x^k = c, \tag{1}$$

$$A_k^T y^k = b_k, \tag{2}$$

$$A_k d^k = a_k. \tag{3}$$

The meaning of the abbreviations b_k, a_k will be clear if we compare with (41) of §12. We observe that the vector a_k will not be known before the system (2) is solved.

In principle, one could solve the three systems (1), (2), and (3) straightforwardly in each exchange step of the simplex algorithm. One could use any of the standard methods (e.g. Gaussian elimination or Householder transformations) to calculate the vectors x^k, y^k, and d^k from (1), (2), and (3) respectively. These and other numerical methods are described in textbooks on numerical analysis, e.g. Dahlquist and Björck

(1974), Stoer (1976) and Stewart (1973).

Such a procedure can make sense in some cases, in particular when the number n is modest, say $n = 10$. However, the computational effort required grows rapidly with n . In a general case it increases as n^3 . Hence the total effort would be prohibitive for problems of a size often encountered in practice, i.e. with hundreds and thousands of variables, even if a large powerful computer is available.

Therefore several variants of the simplex algorithm have been developed in order to reduce the computational labor. The decisive idea is to exploit the fact that the matrices A_{k-1} and A_k are closely related. They differ only by one column vector.

We shall now discuss a variant of the simplex algorithm which is based on Gaussian elimination. The rest of this section is not crucial for the understanding of the simplex algorithm since it deals with the efficient and accurate solution of a sequence of linear systems of equations. Hence the reader may skip this topic during the first reading of the book without losing contact with the contents of succeeding sections.

We consider a linear system of equations of the form

$$Ax = b \tag{4}$$

where $A = (a_{ik})$ $(i,k = 1,\ldots,n)$ is a fixed nonsingular matrix and b a given vector. In order to solve the system of equations one seeks to determine a nonsingular matrix F with the following property:

The product R of F and A ,

$$FA = R \tag{5}$$

is an "upper triangular matrix" of the form

$$R = \begin{pmatrix} r_{11} & r_{12} & \cdots & r_{1n} \\ & r_{22} & \cdots & r_{2n} \\ & & \ddots & \vdots \\ \text{O} & & & r_{nn} \end{pmatrix} \quad \text{with } r_{ii} = 0, \ i = 1,\ldots,n.$$

(5) is called a *triangular factorization* of the matrix A .

(6) <u>The factorization method for linear systems of equations</u>. Suppose a triangular factorization (5) is known. Then the system

$$Ax = b$$

is equivalent to the system

$$Rx = Fb. \tag{7}$$

In order to solve $Ax = b$ one first calculates the vector $\tilde{b} = Fb$ and then solves the system

$$
Rx = \begin{pmatrix}
r_{11}x_1 + r_{12}x_2 + \cdots + r_{1n}x_n \\
\phantom{r_{11}x_1} r_{22}x_2 + + r_{2n}x_n \\
\vdots \\
r_{nn}x_n
\end{pmatrix}
= \begin{pmatrix}
\tilde{b}_1 \\
\tilde{b}_2 \\
\vdots \\
\tilde{b}_n
\end{pmatrix}
= Fb.
$$

The last system is easily solved by means of back-substitution:

$$
\begin{aligned}
x_n &= r_{nn}^{-1}\tilde{b}_n \\
x_{n-1} &= r_{n-1,n-1}^{-1}(\tilde{b}_{n-1} - r_{n-1,n}x_n) \\
\vdots \quad &\quad \vdots \\
x_1 &= r_{11}^{-1}(\tilde{b}_1 - r_{12}x_2 - \cdots - r_{1n}x_n).
\end{aligned}
\tag{8}
$$

(9) Solution of $A^Tx = b$. The system of equations

$$
A^Tx = b \tag{10}
$$

which contains the transpose A^T of A can also be easily solved when a factorization (5) is available. Indeed, (10) is equivalent to the two systems of equations

$$
R^Ty = b \tag{11}
$$

$$
x = F^Ty. \tag{12}
$$

(This statement is verified by multiplying (12) by $A^T = R^T(F^T)^{-1}$.) To solve (10) one starts by determining y from (11):

$$
R^Ty = \begin{pmatrix}
r_{11}y_1 \\
r_{12}y_1 + r_{22}y_2 \\
\vdots \quad \vdots \\
r_{1n}y_1 + r_{2n}y_2 + \cdots + r_{nn}y_n
\end{pmatrix}
= \begin{pmatrix}
b_1 \\
b_2 \\
\vdots \\
b_n
\end{pmatrix}
= b.
$$

Thus y is calculated by means of forward-substitution and one finds $y_1, \ldots, y_n$ in analogy with (8). The solution x is subsequently found from (12) without major effort. Consider now exchange step k of the simplex algorithm. Let a triangular factorization

$$
F_k A_k = R_k
$$

of the basis matrix A_k be known. Then the three linear systems of equations which appear in this exchange step,

$$A_k x^k = c,$$
$$A_k^T y^k = b_k,$$
$$A_k d^k = a_k,$$

may be solved as described in (6) and (9).

(13) <u>Numerical schemes for triangular factorization</u>. The most common methods for calculating a triangular factorization of the type

$$FA = R$$

are based on the following idea. Put

$$A^{(1)} = A$$

and determine a sequence of matrices $A^{(2)},\ldots,A^{(n-1)}$ according to the rules

$$A^{(2)} = F^{(1)} A^{(1)}$$
$$A^{(3)} = F^{(2)} A^{(2)} = F^{(2)} F^{(1)} A$$
$$\vdots$$
$$A^{(n)} = F^{(n-1)} A^{(n-1)} = F^{(n-1)} \ldots F^{(1)} A.$$

Here $F^{(1)},\ldots,F^{(n-1)}$ is another sequence of matrices which are determined such that $A^{(2)},\ldots,A^{(n)}$ take the form indicated below (here "x" means that the element at this point may be different from 0)

$$A^{(2)} = \begin{pmatrix} x & x & \cdots & x \\ 0 & x & \cdots & x \\ 0 & x & \cdots & x \\ \vdots & \vdots & & \vdots \\ 0 & x & & x \end{pmatrix} \qquad A^{(3)} = \begin{pmatrix} x & x & x & \cdots & x \\ 0 & x & x & \cdots & x \\ 0 & 0 & x & \cdots & x \\ 0 & 0 & x & \cdots & x \\ \vdots & \vdots & \vdots & & \vdots \\ 0 & 0 & x & \cdots & x \end{pmatrix}, \ldots$$

$$A^{(n)} = \begin{pmatrix} x & x & x & \cdots & x \\ & x & x & \cdots & x \\ & & x & \cdots & x \\ & \bigcirc & & & \vdots \\ & & & & x \end{pmatrix}. \tag{14}$$

Next we put $A^{(n)} = R$. The triangular factorization sought is then written

$$F^{(n-1)} \ldots F^{(1)} A = R,$$

i.e.

$$FA = R$$

with

$$F = F^{(n-1)} \dots F^{(1)}.$$

Thus the original matrix A is brought to triangular form by means of $n-1$ transformation steps.

Suitable matrices $F^{(1)}, \dots, F^{(n-1)}$ can be calculated in several different ways. We mention here the *Householder* transformations, in which $F^{(i)}$, $i = 1, \dots, n-1$ are orthogonal matrices, and the method based on *Gaussian elimination*. In the latter method one selects

$$F^{(i)} = G_i P_i, \quad i = 1, \dots, n-1,$$

where G_i are so-called elimination matrices and P_i permutation matrices. (See below.) Due to space limitations we shall treat this method only.

(15) <u>Triangular factorization by means of Gaussian elimination.</u> We start by describing the first step of the method (13); i.e. the determination of $F^{(1)}$ such that

$$F^{(1)} A = A^{(2)}.$$

Here $A^{(2)}$ shall have the form (14). We borrow from Gaussian elimination the idea of forming $A^{(2)}$ by subtracting suitable multiples of the first row from the other rows of the matrix A in order to render zero the elements of the first column in the second row, third row, etc. We assume first that

$$a_{11} \neq 0.$$

The following "elimination matrix" has the desired effect:

$$\begin{pmatrix} 1 & & & & & \\ -a_{21}/a_{11} & 1 & & & \mathbf{O} & \\ -a_{31}/a_{11} & 0 & 1 & & & \\ \vdots & & & \ddots & & \\ -a_{n1}/a_{11} & 0 & \dots & 0 & 1 \end{pmatrix}. \tag{16}$$

One verifies this by means of straightforward calculation. If $a_{11} = 0$, one must proceed otherwise and *exchange rows*: one determines an element $a_{i1} \neq 0$ and lets the first and the i-th rows change places. The matrix which results is then multiplied by an elimination matrix (16).

In order to secure numerical stability, it is recommended to choose as the *pivot element* that element in the first row which has the largest absolute value:

$$|a_{i1}| = \max_{k=1,\ldots,n} |a_{k1}|.$$

(17) **Exercise** (**Permutation matrices**). Denote by $\Pi^{(i,k)}$ the $n \times n$ matrix

$$
\begin{array}{l}
\text{Row number i} \rightarrow \\
\\
\\
\text{Row number k} \rightarrow
\end{array}
\left(
\begin{array}{ccccccccc}
1 & & & & & & & & \\
 & \ddots & & & & & & & \\
 & & 1 & & & & & & \\
 & & & 0 & \ldots & 1 & & & \\
 & & & 1 & & & & & \\
 & & & \vdots & \ddots & \vdots & & & \\
 & & & & & 1 & & & \\
 & & & 1 & \ldots & 0 & & & \\
 & & & & & & 1 & & \\
 & & & & & & & \ddots & \\
 & & & & & & & & 1
\end{array}
\right)
$$

Thus we get $\Pi^{(i,k)}$ be interchanging rows number i and k in a unit matrix. Show that $\Pi^{(i,k)}A$ is obtained by exchanging rows number i and k of A. Determine also $A\,\Pi^{(i,k)}$. Finally, show that

$$\Pi^{(i,k)} \cdot \Pi^{(i,k)} = I \quad \text{(unit matrix)}.$$

We have thus constructed a matrix of type (14) by performing one step of the Gaussian elimination process. Hence we obtain

$$A^{(2)} = F^{(1)} A$$

where

$$F^{(1)} = G_1 P_1.$$

Here P_1 is a permutation matrix and G_1 an elimination matrix.

(18) **The general elimination step.** Let the matrix $A^{(k)}$ be given. We now describe how to determine $A^{(k+1)}$ when $A^{(k)}$ is of the form

$$
A^{(k)} = (a_{ij}^{(k)}) =
\left(
\begin{array}{cccccccc}
x & x & x & \cdots & & & x \\
0 & x & & & & & x \\
 & & \ddots & & & & \\
 & & & x & & & \\
 & & & 0 & x & \cdots & \\
 & & & 0 & x & & \\
 & \mathbf{O} & & \vdots & \vdots & & \\
 & & & 0 & x & \cdots & x
\end{array}
\right)
\begin{array}{l}
\\
\\
\\
\\
\leftarrow \text{k-th row}
\end{array}
$$

$$\underset{\uparrow}{\text{k-th column}}$$

We now perform the following operations:

i) Consider the elements in column number k which are on or below the main diagonal of $A^{(k)}$. Determine an element out of these which has largest absolute value. Let $a_{\ell k}^{(k)}$ be such an element, i.e.

$$|a_{\ell k}^{(k)}| = \max_{k \leq i \leq n} |a_{ik}^{(k)}|.$$

ii) Interchange rows number ℓ and k of the matrix $A^{(k)}$, i.e. form the matrix

$$P_k A^{(k)}$$

where $P_k = \Pi^{(\ell, k)}$. (See Exercise (17).) Substeps i) and ii) are often referred to as *row-pivoting*.

iii) Consider row number k in $P_k A$. Subtract suitable multiples of this row from all rows with numbers $k+1, \ldots, n$ in such a manner that all elements in the k-th column and below the main diagonal become zero. This means that we form

$$A^{(k+1)} = G_k A_k$$

where G_k has the form

$$\begin{pmatrix} 1 & & & & & & \\ & \cdot & & & & & \\ & & \cdot & & & O & \\ & & & 1 & & & \\ O & & g_{k+1,k} & 1 & & & \\ & & & \cdot & \cdot & & \\ & & & \cdot & & \cdot & \\ & & g_{nk} & O & & \cdot & 1 \end{pmatrix} \quad \leftarrow \text{k-th row} \qquad (19)$$

with $|g_{ik}| \leq 1$, $i = k+1, \ldots, n$. As a direct consequence of this scheme we get

(20) **Theorem.** Let A be a nonsingular matrix. Then there are permutation matrices $P_1, \ldots, P_{n-1}$ and elimination matrices $G_1, \ldots, G_{n-1}$ such that

$$G_{n-1} P_{n-1} \cdots G_1 P_1 A = R$$

and R is an upper triangular matrix with $r_{ii} \neq 0$, $i = 1, \ldots, n$.

(21) **Numerical realization.** One normally stores the elements $g_{i,k-1}$ of the elimination matrices G_{k-1} in the positions of the matrices

$A^{(k)}$ which are zeroed during the course of the computations. It is also necessary to keep track of the row interchanges. We shall here describe a procedure which is advantageous to use in conjunction with the simplex algorithm, especially when one applies the "stable updating" to be discussed later.

We want to store explicitly the matrix

$$F = G_{n-1} P_{n-1} \cdots G_1 P_1$$

which is obtained by multiplying the $n \times n$ unit matrix by P_1, G_1 $\cdots P_{n-1} G_{n-1}$. This structure is exploited as follows. At the start one stores the $n \times 2n$ matrix

$$B = (A,I).$$

All row operations which are needed for the transfer from $A^{(k)}$ to $A^{(k+1)}$ (row interchanges, additions of multiples of a certain row to other rows) are carried out on all of B. In this way we get the sequence of matrices

$$B^{(1)} = B = (A,I)$$

$$B^{(2)} = G_1 P_1 B = (A^{(2)}, G_1 P_1)$$

$$\vdots$$

$$B^{(n)} = G_{n-1} P_{n-1} B^{(n-1)} = (A^{(n)}, G_{n-1} P_{n-1} \cdots G_1 P_1)$$

$$= (R,F).$$

Thus the matrix B has been replaced by the matrices R and F.

(22) <u>Example</u>. We want to factorize

$$A = \begin{pmatrix} 4 & 2 & 3 \\ 2 & 3 & 1 \\ -3 & 5 & 2 \end{pmatrix}.$$

Thus we put

$$B = (A,I) = \begin{pmatrix} 4 & 2 & 3 & 1 & 0 & 0 \\ 2 & 3 & 1 & 0 & 1 & 0 \\ -3 & 5 & 2 & 0 & 0 & 1 \end{pmatrix} = B^{(1)}.$$

No row interchange is required in the first step since the element in the first column with the largest absolute value is in the first row. According to iii) of (18) we subtract 1/2 times the first row from the second row and (-3/4) times the first row from the third row. We then obtain

$$B^{(2)} = (A^{(2)}, G_1 P_1) = \begin{pmatrix} 4 & 2 & 3 & 1 & 0 & 0 \\ 0 & 2 & -1/2 & -1/2 & 1 & 0 \\ 0 & 13/2 & 17/4 & 3/4 & 0 & 1 \end{pmatrix}.$$

The second and third rows are now interchanged:

$$P_2 G^{(2)} = (P_2 A^{(2)}, \ P_2 G_1 P_1) = \begin{pmatrix} 4 & 2 & 3 & 1 & 0 & 0 \\ 0 & 13/2 & 17/4 & 3/4 & 0 & 1 \\ 0 & 2 & -1/2 & -1/2 & 1 & 0 \end{pmatrix}.$$

The last elimination step (subtraction of a suitable multiple of the second row from the third) gives

$$B^{(3)} = (R,F) = \begin{pmatrix} 4 & 2 & 3 & 1 & 0 & 0 \\ 0 & 13/2 & 17/4 & 3/4 & 0 & 1 \\ 0 & 0 & -47/26 & -19/26 & 1 & -4/13 \end{pmatrix}$$

It is easy to check that

$$FA = \begin{pmatrix} 1 & 0 & 0 \\ 3/4 & 0 & 1 \\ -19/26 & 1 & -4/13 \end{pmatrix} \begin{pmatrix} 4 & 2 & 3 \\ 2 & 3 & 1 \\ -3 & 5 & 2 \end{pmatrix} = \begin{pmatrix} 4 & 2 & 3 \\ 0 & 13/2 & 17/4 \\ 0 & 0 & -47/26 \end{pmatrix} = R.$$

(23) Exercise. Factorize the following matrix

$$A = \begin{pmatrix} 3 & 1 & 6 \\ 2 & 1 & 3 \\ 1 & 1 & 1 \end{pmatrix}.$$

(24) Remarks. (a) The factorization

$$FA = R$$

is closely related to the so-called LR-decomposition of A. Thus one can show that

$$F = G'_{n-1} \ \cdots \ G'_1 \ P$$

where

$$G'_i = P_{n-1} \ \cdots \ P_{i+1} \ G_i \ P_{i+1} \ \cdots \ P_{n-1}, \quad i = 1, \ldots, n-1,$$
$$G'_{n-1} = G_{n-1},$$

and

$$P = P_{n-1} \ \cdots \ P_1.$$

Every G'_i is again a matrix of type (19). (Why?) Therefore

$$G'^{-1}_1 \ G'^{-1}_2 \ \cdots \ G'^{-1}_{n-1} = L$$

is a lower triangular matrix, which is easily verified by means of straight-forward calculation. One obtains the decomposition

$$PA = LR$$

where $\ell_{ii} = 1, \ i = 1, \ldots, n.$

(b) The method for calculating the factorization $FA = R$ which we
have described above is numerically stable with respect to the round-offs
which occur during the course of the computations. We shall not prove this
fact here. This stability is the reason for the use of factorization
methods in "modern" realizations of the simplex algorithm.

Consider now the k-th exchange step of the simplex algorithm. Let
A_k be the basis matrix. The matrices F_k and R_k in the factorization

$$F_k A_k = R_k$$

are calculated as described in (21). The solutions x^k, y^k and d^k to
the linear system of equations

$$A_k x^k = c,$$
$$A_k^T y^k = b_k,$$
$$A_k d^k = a_k$$

are determined as described in (6) and (9). We have already said that one
should exploit the fact that two successive basis matrices A_k and A_{k+1}
differ only in one column in order to get an efficient numerical realization of the simplex method. We now show how to update a factorization of
A_k; i.e. to calculate the factorization of A_{k+1} from that of A_k.

(25) Modification techniques. By this we mean methods which allow
us to pass from a decomposition

$$FA = R$$

of the $n \times n$ matrix A to the corresponding decomposition

$$\tilde{F}\tilde{A} = \tilde{R}$$

where $\tilde{A}$ arises from A as a result of "small changes"; e.g. change of a
row or column or the addition or deletion of a row or column.

Use the same notations as before. Let A be a fixed $n \times n$ matrix
and suppose the decomposition

$$FA = R \qquad\qquad (26)$$

is known. We denote the column vectors of A by $a_1,\ldots,a_n$:

$$A = (a_1,\ldots,a_n).$$

Let a^* be a fixed vector. We consider the matrix $\tilde{A}$ which arises from
A when the r-th column vector is removed from A and the vector a^* is
added as the last column:

$$\tilde{A} = (a_1,\ldots,a_{r-1},\ a_{r+1},\ldots,a_n,a^*).$$

We seek the matrix

$$F\tilde{A} = \tilde{H}.$$

The vectors Fa_i, $i = 1,\ldots,n$ are known from (26) since they are the column vectors of R. Thus

$$\tilde{H} = (Fa_1,\ldots,Fa_{r-1},\ Fa_{r+1},\ldots,Fa_n,\ Fa^*)$$

and $\tilde{H}$ is a matrix of the following form:

$$F\tilde{A} = \tilde{H} = \begin{pmatrix} x & x & \cdots & x & x & \cdots & x \\ & x & & x & x & & x \\ & & & \cdot & \cdot & & \cdot \\ & & & \cdot & \cdot & & \\ & & & x & x & & \\ & O & & & x & & \\ & & & & x & x & \\ & & & & & x & \\ & & & & & & x & x \end{pmatrix}$$

$$\uparrow$$
$$\text{r-th column}$$

The first $r-1$ columns of $\tilde{H}$ are identical with those of R and the r-th through $(n-1)$-th columns of $\tilde{H}$ coincide with the last $n-r$ columns of R. The last column of $\tilde{H}$ is the vector Fa^*.

The matrix $\tilde{H}$ can now be brought into upper triangular form by means of a sequence of Gaussian elimination steps with row-pivoting. (Compare with (18).) Here one needs only to consider the exchange of two neighboring rows. One thus obtains an upper triangular matrix R through

$$\tilde{G}_{n-1}\,\tilde{P}_{n-1}\,\cdots\,\tilde{G}_r\,\tilde{P}_r\,\tilde{H} = \tilde{R}. \tag{27}$$

Each matrix $\tilde{G}_i$ has the form

$$\begin{array}{l} \\ \\ \\ \text{row } i \;\rightarrow \\ \text{row } i+1 \;\rightarrow \end{array} \begin{pmatrix} 1 & & & & & & \\ & 1 & & & O & & \\ & & \cdot & & & & \\ & & & 1 & & & \\ O & & \tilde{g}_i & 1 & & & \\ & & & & \cdot & & \\ & & & & & \cdot & \\ & & & & & & 1 \end{pmatrix}.$$

$\tilde{P}_i$ is either the unit matrix or the matrix which arises from the unit matrix by interchanging rows number i and $i+1$. We have further that

$|g_i| \leq 1$.

From (27) we get

$$\tilde{G}_{n-1}\; \tilde{P}_{n-1}\; \cdots\; \tilde{G}_r\; \tilde{P}_r\; F\tilde{A} = \tilde{R}.$$

Putting

$$\tilde{F} = \tilde{G}_{n-1}\; \cdots\; \tilde{G}_r\; \tilde{P}_r\; F,$$

we get the decomposition sought:

$$\tilde{F}\tilde{A} = \tilde{R}.$$

(28) <u>Numerical realization of the modification</u>. Let the factorization $FA = R$ be calculated according to (21) and given in the form of the
$n \times 2n$ matrix

$$(R,F).$$

One passes to the matrix

$$(\tilde{H},F)$$

where the Hessenberg matrix is formed as described in (25). We now apply
Gaussian elimination to this matrix according to (18) and bring $\tilde{H}$ into
upper triangular form. The final result is then (after $n-r-1$ elimination steps) the matrix

$$(\tilde{R},\tilde{F})$$

and the desired decomposition of $\tilde{A}$ is $\tilde{F}\tilde{A} = \tilde{R}$.

The validity of the procedure just given is a consequence of the
relations

$$\tilde{R} = (\tilde{G}_{n-1}\; \tilde{P}_{n-1}\; \cdots\; \tilde{G}_r\; \tilde{P}_r)\; \tilde{H}$$

and

$$\tilde{F} = (\tilde{G}_{n-1}\; \tilde{P}_{n-1}\; \cdots\; \tilde{G}_r\; \tilde{P}_r)\; F.$$

(29) <u>Example</u>. In (22) we calculated the factorization

$$FA = \begin{pmatrix} 1 & 0 & 0 \\ 3/4 & 0 & 1 \\ -19/26 & 1 & -4/13 \end{pmatrix} \begin{pmatrix} 4 & 2 & 3 \\ 2 & 3 & 1 \\ -3 & 5 & 2 \end{pmatrix} = \begin{pmatrix} 4 & 2 & 3 \\ 0 & 13/2 & 17/4 \\ 0 & 0 & -47/26 \end{pmatrix} = R.$$

Now we want to determine the corresponding decomposition $FA = R$ with

$$\tilde{A} = \begin{pmatrix} 4 & 3 & 1 \\ 2 & 1 & 2 \\ -3 & 2 & 4 \end{pmatrix}.$$

$\tilde{A}$ has arisen from A by replacing the second column by the third and the
third by the new column vector $a^* = (1,2,4)^T$. For simplicity, we work

with 3 decimal places. Thus we start with the matrix

$$(R,F) = \begin{pmatrix} 4 & 2 & 3 & 1 & 0 & 0 \\ 0 & 6.50 & 4.25 & 0.750 & 0 & 1 \\ 0 & 0 & -1.81 & -0.731 & 1 & -0.308 \end{pmatrix}.$$

Entering $Fa^* = (1, 4.75, 0.037)^T$ and following the rules of (28) we get

$$(\tilde{H},F) = \begin{pmatrix} 4 & 3 & 1 & 1 & 0 & 0 \\ 0 & 4.25 & 4.75 & 0.750 & 0 & 1 \\ 0 & -1.81 & 0.037 & -0.731 & 1 & -0.308 \end{pmatrix}.$$

Only one elimination step is required. We must add the second row, multiplied by

$$1.81/4.25 = 0.426$$

to the third row. We then get

$$(\tilde{R},\tilde{F}) = \begin{pmatrix} 4 & 3 & 1 & 1 & 0 & 0 \\ 0 & 4.25 & 4.75 & 0.750 & 0 & 1 \\ 0 & 0 & 2.06 & -0.412 & 1 & 0.118 \end{pmatrix},$$

which defines, within working precision, the factorization $\tilde{F}\tilde{A} = \tilde{R}$.

Check

$$\tilde{F}\tilde{A} = \begin{pmatrix} 1 & 0 & 0 \\ 0.750 & 0 & 1 \\ -0.412 & 1 & 0.118 \end{pmatrix} \begin{pmatrix} 4 & 3 & 1 \\ 2 & 1 & 2 \\ -3 & 2 & 4 \end{pmatrix} = \begin{pmatrix} 4 & 3 & 1 \\ 0 & 4.25 & 4.75 \\ 0.002 & 0.000 & 2.06 \end{pmatrix}.$$

By simply counting the number of multiplications and divisions required, one finds that about $n^3/3$ such operations are necessary to determine the factorization $FA = R$ when A is an $n \times n$ matrix. On the other hand, on the order of magnitude of n^2 multiplications and divisions are needed to carry out one modification. If we neglect the "administrative overhead" of the computational program, the addition and subtractions, we may conclude that the use of modification techniques will entail substantial savings for large n. However, for contemporary computers (1981) it is a very rough approximation to neglect the time required for an addition or subtraction in comparison to that needed for a multiplication or division. In the present case the conclusion will not be altered even if we consider additions and subtractions as well. Normally, the administrative overhead will increase slowly with n. Modification techniques are almost a must for treating large linear optimization problems within reasonable time.

(30) Summary. Let a starting basic matrix A_0 be given. Define the $n \times 2n$ matrix

$$B_0 = (A_0, I)$$

and determine, as described in (21), the matrix

$$(R_0, F_0)$$

such that

$$F_0 R_0 = A_0$$

(n-1 row-pivoting and elimination steps are required). We discuss the
general step. Suppose the matrix

$$B_k = (R_k, \; F_k)$$

has been calculated. The basis matrix A_k has the factorization

$$F_k A_k = R_k.$$

If now the column vector a_r is to be removed and the entering vector $a*$
is determined, one calculates

$$B_{k+1} = (R_{k+1}, \; F_{k+1})$$

as described in (25) and (28). In this way we find the factorization of
the "new" basic matrix A_{k+1}:

$$F_{k+1} A_{k+1} = R_{k+1}.$$

§15.   CALCULATING A BASIC SOLUTION

In order to start the simplex algorithm we need a basic solution
$\{\sigma^0, x^0\}$ of (D). We shall now describe how to construct such a starting
solution.

We consider again the linear optimization problem

(P)   Minimize $c^T y$ subject to $a(s)^T y \geq b(s)$, $s \in S$.

where S is an arbitrary index set.

As in (16) of §11 we introduce the *regularized* problem (F > 0 is a
fixed real number)

(P_F) Minimize $c^T y$ subject to $a(s)^T y \geq b(s)$, $s \in S$

$$e_r^T y \geq -F,$$
$$-e_r^T y \geq -F, \qquad r = 1, \ldots, n.$$

Here, e_r is the r-th unit vector and the last 2n constraints are equivalent to

$$|y_r| \leq F, \quad r = 1,\ldots,n. \tag{1}$$

If (P) has a solution y such that $|y_r| \leq F$ then y is a solution to (P_F) as well and the values of (P) and (P_F) coincide. Hence one can solve (P_F) instead of (P).

(2) <u>The dual problem</u> (D_F). Let the dual variables associated with the last $2n$ constraints of (P_F) be m_r^+, m_r^-, $r = 1,\ldots,n$. The dual takes the form

(D_F) Maximize $\displaystyle\sum_{i=1}^{q} b(s_i)x_i - F \sum_{r=1}^{n} (m_r^+ + m_r^-)$ subject to

$$\sum_{i=1}^{q} a_r(s_i)x_i + m_r^+ - m_r^- = c_r, \quad r = 1,\ldots,n$$

$$x_i \geq 0, \quad i = 1,\ldots,q$$

$$\begin{aligned} m_r^+ &\geq 0, \\ m_r^- &\geq 0, \end{aligned} \quad r = 1,\ldots,n.$$

The second term in the preference function of (D_F),

$$-F \sum_{r=1}^{n} (m_r^+ + m_r^-), \tag{3}$$

may be interpreted as a "penalty" for violating the constraints

$$\sum_{i=1}^{q} a_r(s_i)x_i = c_r, \quad r = 1,\ldots,n.$$

If F is large enough, the constraints will be satisfied.

The advantage of considering the regularized problem stems from the fact that one can immediately find a basic solution of (D_F).

(4) <u>Construction of a basic solution of</u> (D_F). Put $x_i^0 = 0$ and consider the system

$$\overset{o}{m}_r^+ - \overset{o}{m}_r^- = c_r, \quad r = 1,\ldots,n. \tag{5}$$

We get a basic solution of (5) and hence also of (D_F) by putting, for each r, one of the vectors e_r or $(-e_r)$ in the basis. We select e_r if $c_r \geq 0$; otherwise $(-e_r)$ goes into the basis. (5) can now be written as follows:

$$\sum_{r=1}^{n} e_r m_r^+ + \sum_{r=1}^{n} (-e_r) m_r^- = c.$$

We note that the basic solution is *regular* if

$$c_r \neq 0, \quad r = 1,\dots,n.$$

Otherwise it is degenerate. The simplex algorithm can of course be started in both cases.

(6) <u>Remark on the value of the parameter F</u>. The starting method described above can always be used when a suitable a priori estimate of the solutions of (D) is available. If F is chosen too small, however, then the solutions of (D_F) are not feasible for (D). Hence it is not possible to start with the basic solution of (D_F) given in (4) and to use the simplex algorithm to find a basic solution free from all the vectors e_r and $-e_r$, $r = 1,\dots,n$, or with all the corresponding variables m_r^+, m_r^- equal to zero. In this case one could of course increase F and continue with the simplex algorithm.

One arrives to the so-called two-phase method of the simplex procedure by arguing as follows. If no "realistic" estimate (1) is available then one chooses F very large. This means that the first term of the preference function of (D_F) has a relatively small influence. It can therefore be neglected. Then we consider instead the problem:

$$\text{Maximize} \quad -\sum_{r=1}^{n} (m_r^+ + m_r^-) \tag{7}$$

$$\text{subject to the constraints of } (D_F).$$

(8) <u>Phase I of the simplex procedure</u>. The simplex algorithm is applied to the following dual pair of linear optimization problems:

(P_I) Minimize $c^T y$ subject to $a(s)^T y \geq 0$, $s \in S$

$$e_r^T y \geq -1,$$
$$\qquad\qquad\qquad\qquad r = 1,\dots,n.$$
$$-e_r^T y \geq -1,$$

(D_I) Maximize $-\sum_{r=1}^{n} (m_r^+ + m_r^-)$ subject to the constraints of (D_F).

(9) <u>Exercise</u>. Confirm that (P_I) and (D_I) form a dual pair of linear optimization problems. Also discuss in what sense (P_I) can be looked upon as a limiting case of (P_F) when $F \to \infty$.

When (D_I) is treated with the simplex algorithm one seeks to satisfy the constraints of (D). If $v(D_I) = 0$, then (D) is consistent, and if $v(D_I) < 0$, then (D) is inconsistent.

We assume now that the simplex algorithm has delivered an optimal basic solution to (D_I) after finitely many exchange steps and that the corresponding value of the dual preference function is zero. Thus (D) is feasible.

The basis vectors of this optimal basic solution are called a_i, $i = 1,\ldots,n$:

$$a_i \in \{a(s) \mid s \in S\} \cup \{e_r \mid r = 1,\ldots,n\} \cup \{-e_r \mid r = 1,\ldots,n\}.$$

If now all a_i are of the form

$$a_i = a(s_i), \quad i = 1,\ldots,n,$$

i.e. *none* of the vectors e_r and $-e_r$ appear in the basis, then one may put

$$\sigma^0 = \{s_1,\ldots,s_n\},$$

which is then a basic set for (P) - (D). We may now apply the simplex algorithm to (P) - (D) and have thus entered Phase II of the simplex procedure. Hence this is always possible if the optimal value of (D_I) is zero *and* Phase I of the simplex procedure delivers a *regular* optimal basic solution. If the value of (D_I) is zero, then none of the vectors e_r or $(-e_r)$ can appear in the optimal basis with a positive weight m_r^+ or m_r^- respectively.

We assume now that (D_I) has the optimal value zero and a degenerate optimal basic solution where at least one of the vectors a_i is of the form e_r or $-e_r$. Then one proceeds in Phase II of the simplex method as follows: consider the *modified dual problem:*

(D_{II}) Maximize $\displaystyle\sum_{i=1}^{q} b(s_i)x_i$ subject to

$$\sum_{i=1}^{q} a_r(s_i)x_i + m_r^+ - m_r^- = c_r, \quad r = 1,\ldots,n$$

$$\sum_{r=1}^{n} (m_r^+ + m_r^-) = 0$$

$$x_i \geq 0, \quad i = 1,\ldots,q$$

$$m_r^+ \geq 0, \quad m_r^- \geq 0, \quad r = 1,\ldots,n.$$

The constraint $\Sigma_{r=1}^{n} (m_r^+ + m_r^-) = 0$ has been introduced to insure that every feasible solution of (D_{II}) satisfies $m_r^+ = 0$, $m_r^- = 0$, $r = 1,\ldots,n$. Therefore D_{II} is equivalent to D. Every basic matrix of D_{II} has, of course, $n+1$ column vectors. We now show how to construct a starting basic set for D_{II} from an optimal basic solution of D_I.

Let an optimal basic matrix of (D_I) contain the following n column vectors:

$$a_1 = a(s_1),\ldots,a_k = a(s_k),$$

$$a_{k+1} = e_{i_1},\ldots,a_{k+\ell} = e_{i_\ell},$$

$$a_{k+\ell+1} = -e_{i_{\ell+1}},\ldots,a_n = -e_{i_{n-k}},$$

where k and ℓ are fixed integers. The basic matrix thus has the form

$$
A = \begin{pmatrix}
\uparrow & & \uparrow & \begin{matrix} \cdot & 0 \\ \cdot & 1 \\ \cdot & 0 \\ 0 & \cdot \\ 1 & \cdot \\ 0 & \cdot \\ & & \\ \cdot & \cdot \\ \cdot & \cdot \end{matrix} & \begin{matrix} \cdot & \cdot \\ \cdot & \cdot \\ \cdot & 0 \\ \cdot & -1 \\ \cdot & 0 \\ & \\ 0 & \\ -1 & \\ 00 & \cdot \end{matrix} \\
a(s_1)\ldots a(s_k) & & & & \\
\downarrow & & \downarrow & &
\end{pmatrix}.
$$

Now let e be any vector out of the set $\{a_{k+1},\ldots,a_n\}$. Then $-e$ *cannot* belong to the set $\{a_{k+\ell+1},\ldots,a_n\}$, since the basic matrix is nonsingular, and the following $(n+1) \times (n+1)$ matrix is a starting basic matrix for (D_{II}):

$$
\hat{A} = \begin{pmatrix} A & & -e \\ 0\ldots0\ 1\ldots1 & & 1 \end{pmatrix}.
$$
$$\underset{\text{k-th column}}{\uparrow}$$

(10) **Exercise.** Show that $\hat{A}$ is nonsingular.

If the calculations are carried out as described in §14, then we leave Phase I with the basic matrix A factorized according to

$$FA = R. \tag{11}$$

This decomposition is recorded as the $n \times 2n$ matrix

$$(R,F).$$

We describe next how to find a corresponding factorization

$$\hat{F}\hat{A} = \hat{R},$$

i.e. the $(n+1) \times 2(n+1)$ matrix

$$(\hat{R},\hat{F})$$

from (R,F). This is done by means of a method similar to the modification techniques of §14.

We find from (11) that

$$\left(\begin{array}{c|c} F & 0 \\ \hline 0 & 1 \end{array}\right)' \cdot \hat{A} = \left(\begin{array}{c|c} R & -Fe \\ \hline 0\ldots0\ 1\ldots1 & 1 \end{array}\right).$$

The matrix on the right is "almost" in triangular form. One needs at most $n-k$ permutation and elimination steps to bring it into triangular form. Consider the $(n+1) \times 2(n+1)$ matrix

$$\left(\begin{array}{c|c|c|c} R & -Fe & F & \begin{array}{c}0\\ \vdots \\ 0\end{array} \\ \hline 0\ldots0\ 1\ldots1 & 1 & 0\ldots0 & 1 \end{array}\right). \tag{12}$$

After $n-k$ Gaussian elimination steps, (12) is changed to assume the form

$$\left(\begin{array}{cccc|ccc} x & x & \cdots & x & x & \cdots & x \\ & x & & & \cdot & & \cdot \\ & & \cdot & & \cdot & & \cdot \\ \mathbf{O} & & & \cdot & \cdot & & \cdot \\ & & & x & x & \cdots & x \end{array}\right) = (\hat{R}, \hat{F}).$$

Hence we have the desired factorization

$$\hat{F}\hat{A} = \hat{R}$$

of the starting basis $\hat{A}$.

(13) **Remark**. It may be more practical to work with an $(n+1) \times 2(n+1)$ matrix already in Phase I. We form it as follows, where A_0 is an $n \times n$ basis matrix:

$$\left(\begin{array}{c|c|c|c} A_0 & 0 & I & 0 \\ \hline 0 & 0 & 0\ldots0 & 0 \end{array}\right)$$

After factorizations we get

$$\left(\begin{array}{cccc} R & 0 & F & 0 \\ 0 & 0 & 0 & 0 \end{array}\right). \tag{14}$$

It is now easy to supplement (14) to obtain (12). Then the matrix

$$(\hat{R}, \hat{F})$$

is calculated as earlier described and one may enter Phase II.

Chapter VII
A General Three-Phase Algorithm

In this chapter we shall describe a computational scheme for efficient numerical treatment of general linear optimization problems with *infinitely* many constraints. For this purpose we shall derive a nonlinear system of equations from whose solutions one constructs an optimal solution of the original problem. The general scheme is then presented and its use is illustrated in several numerical examples.

Thus we consider again the dual pair (P) - (D):

(P)   Minimize $c^T y$ subject to $\sum_{r=1}^{n} a_r(s) y_r \geq b(s)$, $s \in S$.

(D)   Maximize $\sum_{i=1}^{q} b(s_i) x_i$ subject to

$$\sum_{i=1}^{q} a_r(s_i) x_i = c_r, \quad r = 1, 2, \ldots, n,$$

$$x_i \geq 0,$$
$$s_i \in S, \qquad i = 1, \ldots, q.$$

In this chapter we shall require that (P) and (D) are solvable and that no duality gap occurs. We shall further assume that S is a nonempty compact subset of R^k and that $a_1, \ldots, a_n$, b are continuous functions on S. Later, we shall also impose the condition that they have continuous partial derivatives up to a certain order.

§16. NONLINEAR SYSTEMS DERIVED FROM OPTIMALITY CONDITIONS

(1) Theorem. Let y be an optimal solution to Problem (P) and let
$\{s_1,\ldots,s_q;\ x_1,\ldots,x_q\}$ be an optimal solution to (D) with $1 \leq q \leq n$
and such that

$$x_i > 0, \quad i = 1,\ldots,q. \tag{2}$$

Put

$$f(s) = \sum_{r=1}^{n} a_r(s)y_r - b(s).$$

Then $y_1,\ldots,y_n,\ s_1,\ldots,s_q,\ x_1,\ldots,x_q$ have the properties (3), (4) and
(5) below;

$$\sum_{r=1}^{n} a_r(s_i)y_r = b(s_i), \quad i = 1,\ldots,q; \tag{3}$$

$$\sum_{i=1}^{q} a_r(s_i)x_i = c_r, \quad r = 1,\ldots,n. \tag{4}$$

The function f assumes its *minimal value* at $s_1,\ldots,s_q$. $\tag{5}$

Proof: (3) follows from (2) and the duality slackness conditions
(14) of §12. (4) expresses the fact that $\{s_1,\ldots,s_q,\ x_1,\ldots,x_q\}$ is
feasible for (D). Since y is a feasible vector for (P) we have

$$f(s) \geq 0, \quad s \in S. \tag{6}$$

By (3), $f(s_i) = 0, i = 1,\ldots,q$, establishing (5).

In the computational scheme to be described in this chapter, (3), (4),
and (5) will be used for the calculation of $y_1,\ldots,y_n,\ s_1,\ldots,s_q,$
$x_1,\ldots,x_q$. We shall assume that q can be determined, e.g. from a sufficiently fine discretization of (P). We shall call $x_1,\ldots,x_q$ *masses*
and refer to $s_1,\ldots,s_q$ as the corresponding *mass-points*. Thus q is
the *number* of mass-points.

(7) Remark. Since we have assumed that S is a subset of R^k,
each mass-point s_i corresponds to a vector with k components
$s_i^1,\ldots,s_i^k$, $(i = 1,\ldots,q)$. Thus

$$n + kq + q$$

"unknowns" $y_1,\ldots,y_n,\ s_1^1,\ldots,s_1^k,\ s_q^1,\ldots,s_q^k,\ x_1,\ldots,x_q$ will appear in
the calculation of the primal and dual solutions. (3) and (4) will give

$$q + n$$

equations which must be satisfied by these unknowns. The "missing" kq
equations will be derived from (5). Then we will get a system of equations with the same number of equations and unknowns. Its solution is
then used to construct optimal solutions to (P) and (D).

(8) Exercise. Show that $y_1,\ldots,y_n$, $s_1,\ldots,s_q$, $x_1,\ldots,x_q$ are solutions to (P) and (D), if they satisfy (2) - (5).   .

(9) Example. Let the functions a_r, r = 1,...,n, and b have continuous partial derivatives of first order on S. Assume also that s_i,
i = 1,...,q, lie in the interior of S. Then (5) entails

$$\frac{\partial f}{\partial s^j}(s_i) = 0, \quad j = 1,\ldots,k \quad \text{and} \quad i = 1,\ldots,q.$$

Therefore we get in this case the following system of kq equations

$$\sum_{r=1}^{n} \nabla a_r(s_i)y_r = \nabla b(s_i), \quad i = 1,\ldots,q. \tag{10}$$

(The gradient vector ∇f of a real-valued differentiable function f
is here defined by

$$\nabla f(s) = \left(\frac{\partial f}{\partial s^1}(s),\ldots,\frac{\partial f}{\partial s^k}(s)\right)^T .)$$

Thus we get, by combining (3), (4), and (10), a nonlinear system of equations with n + (k+1)q unknowns and the same number of equations. This
system may be treated by means of one of the standard numerical schemes,
e.g. the Newton-Raphson method. See e.g. Dahlquist-Björck (1974) or Stoer
(1976).

(11) Remark. It is well-known that the conditions (10) are necessary for (5) but not sufficient. Thus a solution to (3), (4) and (10)
that also satisfies (2) may not satisfy (5) and hence may not be a solution to (P) and (D). In order to establish that a candidate solution obtained from the necessary conditions really solves the dual pair (P) -
(D), one must verify that the infinitely many primal conditions

$$\sum_{r=1}^{n} a_r(s)y_r \geq b(s), \quad s \in S,$$

are met.

Before we discuss the case when a mass-point is situated on the
boundary of S we shall make some important observations about the
determination of the integer q by means of discretization.

Let (P_ℓ) be a discretization of P (compare §13) and let $\{\sigma_\ell, x_\ell\}$ be an optimal basic solution of the corresponding dual problem (D_ℓ);

$$\sigma_\ell = \{s_{1\ell}, \ldots, s_{n\ell}\} \subset S_\ell,$$
$$x_\ell = \{x_{1\ell}, \ldots, x_{n\ell}\}^T \in R^n.$$

Denote by q_ℓ the number of *positive* components of x_ℓ. We recall that the basic solution is termed *degenerate* if $q_\ell < n$. It seems reasonable to expect that if the grid S_ℓ is sufficiently fine then $q_\ell = q$ where q is the number of masspoints of a solution of (D).

However, several numerical examples have been solved where the contrary is true. In almost all problems, one finds that

$$q_\ell = n$$

for *all* discretizations $(P_\ell) - (D_\ell)$, irrespective of the fineness of the grid. Thus the discretized problems are generally not degenerate. This observation agrees with the theoretical result that the degenerate linear programs are, in a certain sense, more rare than the regular ones. On the other hand, the case

$$q < n$$

is fairly common in optimization problem with infinitely many constraints. Nevertheless, the integer q can be determined from the solutions of discretized problems, as is illustrated in the following example.

(12) **Example.** We want to solve the (primal) problem

$$\text{Minimize} \quad \sum_{r=1}^{8} y_r/r$$

subject to

$$\sum_{r=1}^{8} s^{r-1} y_r \geq 1/(2-s), \quad s \in [0,1] = S.$$

We discretize and select the following subsets S_ℓ of S $(\ell \geq 2)$:

$$S_\ell = \{0, h_\ell, 2h_\ell, \ldots, 1\} \quad \text{with} \quad h_\ell = 1/(\ell - 1).$$

The corresponding discretized Problems $(P_\ell) - (D_\ell)$ were solved on a computer by means of the simplex algorithm for $\ell = 21, 41, 81$. $q_\ell = 8$ was obtained in all three cases and the following basic sets emerged:

$\ell = 21$	$\ell = 41$	$\ell = 81$	Group
σ_{21}	σ_{41}	σ_{81}	
0.0000	0.0000	0.0000	1
0.1500	0.1500	0.1625	2
0.2000	0.1750	0.1750	
0.5000	0.5000	0.5000	3
0.5500	0.5250	0.5125	
0.8000	0.8250	0.8250	4
0.8500	0.8500	0.8375	
1.0000	1.0000	1.0000	5

We note that the eight numbers in each column may be divided into five groups. The elements of Groups number 2,3,4 lie closely together and the distances between the two elements in these groups get smaller with increasing ℓ.

It is reasonable to assume that

$$q = 5$$

holds for the "continuous" problem. This conjecture can be shown to be true by means of the theory of Chapter VIII. Now we shall demonstrate how to derive a nonlinear system for the primal and dual problems of this particular example. There are 8 primal unknowns, namely $y_1, \ldots, y_8$, and $q = 5$ mass-points $s_1, \ldots, s_5$ with the corresponding masses $x_1, \ldots, x_5$. The results of the table above indicate $s_1 = 0$ and $s_5 = 1$, i.e. we assume that these points lie on the boundary of the interval $[0,1]$. This may also be concluded from the theory of Chapter VIII. There remain the 8 dual unknowns

$$s_2, s_3, s_4, x_1, x_2, x_3, x_4, x_5.$$

Hence there are in total 8 unknown numbers to determine. Now $q = 5$ equations result from (3) and $n = 8$ equations from (4). The missing 3 equations are obtained from (5) and the observation that the "error function" f assumes its minimum value at s_2, s_3, s_4. Hence its derivative must vanish at these points, giving the 3 equations sought. We give now the 16 equations (observe that $s_1 = 0$ and $s_5 = 1$):

$$y_1 \qquad\qquad\qquad - 1/2 \qquad = 0$$

$$y_1 + s_2 y_2 + s_2^2 y_3 + \ldots + s_2^7 y_8 - 1/(2-s_2) = 0$$

$$y_1 + s_3 y_2 + s_3^2 y_3 + \ldots + s_3^7 y_8 - 1/(2-s_3) = 0$$

$$y_1 + s_4 y_2 + s_4^2 y_3 + \ldots + s_4^7 y_8 - 1/(2-s_4) = 0$$

$$y_1 + y_2 + y_3 + \ldots + y_8 - 1 \qquad = 0$$

$$x_1 + x_2 + x_3 + x_4 + x_5 - 1 \qquad = 0$$

$$s_2 x_2 + s_3 x_3 + s_4 x_4 + x_5 - 1/2 = 0$$

$$s_2^2 x_2 + s_3^2 x_3 + s_4^2 x_4 + x_5 - 1/3 = 0$$

$$\vdots \qquad\qquad \vdots \qquad\qquad \vdots$$

$$s_2^7 x_2 + s_3^7 x_3 + s_4^7 x_4 + x_5 - 1/8 = 0$$

$$y_2 + 2 s_2 y_3 + \ldots + 7 s_2^6 y_8 - 1/(2-s_2)^2 = 0$$

$$y_2 + 2 s_3 y_3 + \ldots + 7 s_3^6 y_8 - 1/(2-s_3)^2 = 0$$

$$y_2 + 2 s_4 y_3 + \ldots + 7 s_4^6 y_8 - 1/(2-s_4)^2 = 0.$$

We recommend that the reader verify each of these equations. (The first equation gives immediately $y_1 = 1/2$ and this value can be entered into the remaining equations decreasing the size of the system somewhat.) We write the above system in the form

$$q_1(y,s,x) = 0$$
$$\vdots \qquad\qquad\qquad\qquad\qquad (13)$$
$$q_{16}(y,s,x) = 0$$

where we use the notation

$$(y,s,x) = (y_1,\ldots,y_8,s_2,s_3,s_4,\ x_1,\ldots,x_5).$$

We next show how to construct an *approximate* solution $(\bar{y},\bar{s},\bar{x})$ to (13) from the solutions $(y_\ell;\sigma_\ell,x_\ell)$ of $(P_\ell) - (D_\ell)$. The approximation $(\bar{y},\bar{s},\bar{x})$ may then be improved by an iterative scheme, e.g. the Newton-Raphson method. We put

$$\bar{y}_{i\ell} = y_{i\ell},\ i = 1,2,\ldots,8 \quad (\text{here, } y_\ell = (y_{1\ell},\ldots,y_{8\ell}));$$

$$\bar{x}_{1\ell} = x_{1\ell};\ \bar{x}_{i\ell} = x_{2i-2,\ell} + x_{2i-1,\ell},\ i = 2,3,4;\ \bar{x}_{5\ell} = x_{8\ell}$$

$$\bar{s}_{i\ell} = (x_{2i-2,\ell} s_{2i-2,\ell} + x_{2i-1,\ell} \cdot s_{2i-1,\ell})/\bar{x}_{i\ell},\ i = 2,3,4.$$

Thus $\bar{s}_{i\ell}$ is the *center of gravity* of the mass-points belonging to Group number i.

The 'goodness' of this approximation is expressed by the number

$$\rho_\ell = \max_{i=1,\ldots,16} |q_i(y_\ell, s_\ell, x_\ell)|.$$

Thus ρ_ℓ is the maximum norm of the residual vector of the system (13). We get the following table:

Group	$\ell=21$		$\ell=41$		$\ell=81$	
	$\bar{s}_{i1}$	$\bar{x}_{i1}$	$\bar{s}_{i1}$	$\bar{x}_{i1}$	$\bar{s}_{i1}$	$\bar{x}_{i1}$
1	0	0.048495	0	0.049853	0	0.049940
2	0.174643	0.277444	0.172802	0.272652	0.172742	0.272418
3	0.500004	0.348128	0.500003	0.354998	0.500004	0.355290
4	0.825362	0.277439	0.827202	0.272648	0.827262	0.272413
5	1	0.048494	1	0.049852	1	0.049938
Residual norm ρ_1	$1.40 \cdot 10^{-3}$		$1.16 \cdot 10^{-4}$		$5.22 \cdot 10^{-5}$	

The exact mass-points and masses are given below:

$s_1 = 0$ $\qquad x_1 = 1/20\ = 0.05$

$s_2 = 0.5(1-\sqrt{3/7}) = 0.172673$ $x_2 = 49/180 = 0.272222$

$s_3 = 0.5$ $\qquad x_3 = 16/45\ = 0.355556$

$s_4 = 0.5(1+\sqrt{3/7}) = 0.827327$ $x_4 = 49/180 = 0.272222$

$s_5 = 1$ $\qquad x_5 = 1/20\ = 0.05.$

It is generally true that very good approximate solutions to the nonlinear system of equations can be constructed by means of discretization, linear programming and clustering of mass-points by determining centers of gravity as described above. Sometimes it is not even necessary to improve upon this approximate solution by means of iterative methods.

(14) Exercise. Use the same method as in the preceding example to solve the problem

$$\text{Minimize } \sum_{r=1}^{10} y_r/r \text{ subject to } \sum_{r=1}^{10} s^{r-1} y_r \geq -1/(1+s^2), \quad s \in [0,1] = S.$$

Use for the discretization an equidistant grid with N = 41 points. Solve the corresponding linear program with the simplex algorithm. The approxi-

mate solution found in this way may be compared with the true result which
is as follows: $q = 6$; the mass-points and masses are

i	s_i	x_i
1	0.037989	0.096417
2	0.190708	0.202986
3	0.427197	0.259604
4	0.686634	0.247041
5	0.897894	0.165077
6	1.000000	0.028876 .

The primal solution is

$$y = (-1.000000, -5.837 \cdot 10^{-5}, 1001582, -0.020238, -0.856457, -0.612559,$$
$$2.622486, -2.606125, 1.188151, -0.216783)^T.$$

In the next section we shall describe a general computational scheme.

§17. A GENERAL COMPUTATIONAL SCHEME

Retain the general assumptions of the beginning of this chapter, in-
cluding the requirement that $a_1, a_2, \ldots, a_n$, and b have continuous partial
derivatives of the first and second order. We propose

(1) A general computational scheme consisting of the three phases
i), ii) and iii) below.

i) The dual pair (P) - (D) is discretized; i.e. the infinite index
set S is replaced by a finite subset. The resulting dual pair of linear
programs is solved by means of the simplex method.

ii) The structure of the nonlinear system (3), (4), (5) of §16 is
determined from the calculated optimal solutions of the discretized problems. A tolerance ε is selected. If among the mass-points of the solution of the discretized problem there are two mass-points s_i and s_j
with masses x_i and x_j such that the distance between s_i and s_j is
less than ε, then they are replaced by a mass-point $\bar{s}$ carrying mass $\bar{x}$
where

$$\bar{x} = x_i + x_j, \quad \bar{x} = (x_i s_i + x_j s_j)/\bar{x}.$$

This procedure is repeated as long as there still are two mass-points lying
closer to each other than ε.

A nonlinear system is now derived by combining (3), (4), (5) of §16.

iii) The nonlinear system obtained in Phase ii) is solved by some numerical procedure, e.g. the Newton-Raphson method. If the calculated solution satisfies the feasibility conditions of (P) and (D), it is accepted as optimal. Otherwise one reenters Phase i) with a refined grid.

Remark. The scheme described above has been successfully applied to many practical problems. It is recommended to use a numerically stable realization of the simplex algorithm in Phase i), e.g. the version described in §14, which uses stable updating of the basic matrix.

In Phase ii) we construct a nonlinear system by combining (3), (4), and (5) of §16. Thus if s_i is an interior point of S, we get k equations from (5) of §16 as explained in (9) of §16. If s_i is a *boundary* point of S, one may proceed as explained in the Example (3) below if S has a simple structure. A more general description can be formulated by means of the so-called Kuhn-Tucker conditions if S is defined through a set of inequalities:

$$S = \{s \in R^k : h_j(s) \leq 0, \quad j = 1,\ldots,p\}.$$

The reader is referred to Collatz and Wetterling (1971) for a discussion of this topic.

(2) Remark. If the tolerance ε is selected too large in Phase ii) or the grid of Phase i) is not sufficiently fine then we may enter Phase iii) with the wrong nonlinear system and the Newton-Raphson iterations diverge or converge to a "solution" which does not define a feasible vector y of (P). In both cases one reenters Phase i) with a finer grid and reduces the tolerance ε in Phase ii). It is possible to show that Phase iii) succeeds provided that the grid in Phase i) is sufficiently fine, ε in Phase ii) is small enough, and certain general regularity conditions are met. A general three-phase scheme for semi-infinite programs of the type given above was first published in Gustafson (1970).

(3) Example. Let S be the set $[0,1] \times [0,1]$, i.e. the unit square in the plane (k = 2). We consider the case n = 8. Assume that after carrying out Phase i) (solution of the discretized version of the dual pair (P) - (D)) we get a mass-point distribution as depicted in Fig. 17.1. 8 mass-points appear in 4 clusters containing 1, 2, 2, and 3 mass-points respectively. Assume now that ε is chosen such that Phase ii) replaces each cluster with one mass-point. Thus q = 4. Hence we must determine the 4 mass-points $s_1,\ldots,s_4$, the corresponding masses $x_1,\ldots,x_4$, and the 8 primal variables $y_1,\ldots,y_8$. Since each s_i is a two-dimensional

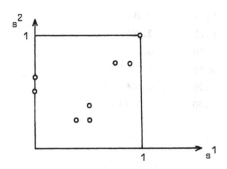

Fig. 17.1

vector we have in total $4 \cdot 2 + 4 + 8 = 20$ unknowns. Due to the character of the solution of the discretized problem we assume that s_1 is in the upper right corner, s_2 on the left-hand boundary and s_3, s_4 in the interior of S. Thus we should have $s_1^1 = s_1^2 = 1$ and $s_2^1 = 0$ where s_1^1 and s_1^2 are the first and second components of s_1. Therefore the total of remaining unknowns is 17. (5) of §16 now gives the 5 equations

$$\sum_{r=1}^{8} \frac{\partial}{\partial s^2} a_r(s_2) y_r - \frac{\partial}{\partial s^2} b(s_2) = 0$$

$$\sum_{r=1}^{8} \frac{\partial}{\partial s^i} a_r(s_j) y_r - \frac{\partial}{\partial s^i} b(s_j) = 0, \quad i = 1,2 \quad \text{and} \quad j = 3,4,$$

with (3) and (4) of §16 giving the remaining 4 and 8 equations. Hence we have constructed a nonlinear system of equations with the same number of equations as unknowns.

(4) Exercise. Describe how to construct the nonlinear system of equations when S is the unit circle in the plane and some mass-points are situated on the boundary.

(5) Example. We consider the first example which was discussed in (13) of §13. Upon discretization we get the mass-points indicated in the table below:

i	Coord. of s_i		x_i
1	1.00	0.00	0.0667
2	0.50	0.25	0.2667
3	0.25	0.50	0.2667
4	0.50	0.50	0.2000
5	0.00	1.00	0.0667
6	1.00	1.00	0.1333

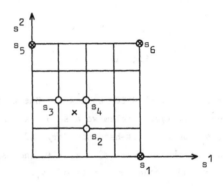

Fig. 17.2

(o Mass-points of the discr. problem)

(x Mass-points of the cont. problem)

Here the 6 mass-points appear in 4 clusters, one of which has 3 members, the other 3 having one mass-point each. Phase ii) gives $q = 4$ and the following initial approximation for the solution to the dual problem (D):

i	Coord. of mass-point s_i		Mass x_i
1	1	0	0.0667
2	0.4091	0.4091	0.7334
3	0	1	0.0667
4	1	1	0.1333

Derive the corresponding nonlinear system of equations (it has 6+4+2 unknowns) and verify that it is satisfied by the optimal solution of the primal problem (given in (13) of §13) together with the following masses and mass-points:

i	Coord. of mass-point i		Mass x_i
1	1	0	0.083333
2	0.400000	0.400000	0.694444
3	0	1	0.083333
4	1	1	0.138888 .

We shall also discuss the class of problems which were treated in §6, namely calculation of uniform approximations. The computational scheme which is described in this chapter has been very efficient for the solution of this class of problems, in particular by approximation on multiple-dimensional sets.

Let T be a compact subset of R^k ($k \geq 1$) with at least $n+1$ elements and let $v_1,\ldots,v_n$, and f be real-valued functions on T. The approximation problem reads

(PA)   Minimize y_{n+1} subject to $\left| \sum_{r=1}^{n} y_r v_r(t) - f(t) \right| \leq y_{n+1}$, $t \in T$.

The corresponding dual becomes

(DA)   Maximize $\sum_{i=1}^{q} f(t_i)x_i$ subject to

$$\sum_{i=1}^{q} v_r(t_i)x_i = 0, \quad r = 1,\ldots,n,$$

$$\sum_{i=1}^{q} |x_i| \leq 1,$$

$$\{t_1,\ldots,t_q\} \subset T.$$

(6) **Theorem.** Let the functions $v_1,\ldots,v_n$ be linearly independent. Then both the problems (PA) and (DA) are solvable and have a joint optimal value. There are always optimal solutions for (DA) such that $x_1 \neq 0,\ldots,x_q \neq 0$ with

$1 \leq q \leq n+1$.

Proof: The proof is obtained by combining (15) of §6 with (12) of §10 and (12) of §11.

The next theorem corresponds to (1) of §16 and its converse (see (8) of §16).

(7) **Theorem.** Let the $n+1$ functions $v_1,\ldots,v_n$, f be linearly independent. Then $y_1,\ldots,y_n$, y_{n+1} with $y_{n+1} \geq 0$ and $\{t_1,\ldots,t_q,$

$x_1,\ldots,x_q\}$ with $t_i \in T$, $x_i \neq 0$, $i = 1,\ldots,q$ $(1 \leq q \leq n+1)$ are optimal for (PA) and (DA) if and only if the following relations hold:

$$f(t_i) - \sum_{r=1}^{n} y_r v_r(t_i) = y_{n+1} \, \mathrm{sgn} \, x_i, \quad i = 1,\ldots,q, \qquad (8)$$

$$\sum_{i=1}^{q} v_r(t_i) x_i = 0, \qquad\qquad r = 1,\ldots,n, \qquad (9)$$

$$\sum_{i=1}^{q} |x_i| = 1. \qquad\qquad (10)$$

The "error function" $\displaystyle\sum_{r=1}^{n} y_r v_r - f$ $\qquad\qquad (11)$

assumes its maximum or minimum value on T at each point $t_1,\ldots,t_q$.

Proof: Consider the linear optimization problem equivalent to (PA). Then the theorem is a direct consequence of (1) and (8) of §16. It is even easy to show that one direction of the statement follows from Lemma (29) of §6. We verify here how (8) is derived from the optimality of $y_1,\ldots,y_n$ and $\{t_1,\ldots,t_q,x_1,\ldots,x_q\}$. If $x_i > 0$ we write, as in the proof of Lemma (15) of §6, $x_i^+ = x_i$, $t_i^+ = t_i$. The corresponding complementary slackness condition is

$$x_i^+\left(\sum_{r=1}^{n} y_r v_r(t_i^+) + y_{n+1} - f(t_i^+)\right) = 0.$$

If $x_i < 0$ we put $x_i^- = -x_i$, $t_i^- = t_i$ and get the complementary slackness condition

$$x_i^-\left(-\sum_{r=1}^{n} y_r v_r(t_i^-) + y_{n+1} + f(t_i^-)\right) = 0.$$

Thus if $x_i \neq 0$ one of the following two equations is satisfied:

$$f(t_i) - \sum_{r=1}^{n} y_r v_r(t_i) = y_{n+1}, \quad \text{if } x_i > 0;$$

$$f(t_i) - \sum_{r=1}^{n} y_r v_r(t_i) = -y_{n+1}, \quad \text{if } x_i < 0.$$

This is equivalent to (8). We point out that the inequality $y_{n+1} > 0$ is a consequence of the linear independence of $v_1,\ldots,v_n,f$.

The numerical treatment of the dual pair (PA) - (DA) is analogous to that of (P) - (D). Thus a three-phase computational scheme is used. The problem is discretized and an initial approximate solution is constructed, giving q. In particular, one observes the sign of an x_i resulting from the solution of the discretized problem. This is taken into account in the equations (8) and (11). As an example consider the case $q = 2$, $x_1 > 0$ and $x_2 < 0$. Then (8) and (10) give the equations

$$f(t_1) - \sum_{r=1}^{n} y_r v_r(t_1) = y_{n+1},$$

$$f(t_2) - \sum_{r=1}^{n} y_r v_r(t_2) = -y_{n+1},$$

$$x_1 - x_2 = 1.$$

The conditions (11) give rise to equations in the same way as (5) of §16. Thus one determines from the results of the discretized problem whether a mass-point t_1 (in this case an extremal point of the error function) lies in the interior or at the boundary of T. Accordingly, one appends conditions that partial derivatives must vanish at t_i.

(12) **Example**. The following problem is treated in Andreasson and Watson (1976). The function

$$f(s,t) = \exp(-s^2 - t^2)$$

is to be approximated in the uniform norm on the square $0 \le s,t \le 1$ by a linear combination of the functions

$$v_1(s,t) = 1, \quad v_2(s,t) = s, \quad v_3(s,t) = t, \quad v_4(s,t) = 2s^2-1$$
$$v_5(s,t) = st, \quad v_6(s,t) = 2t^2-1.$$

At first the problem is discretized; i.e. it is approximated by the task

$$\text{Determine} \quad \min_{y_1,\ldots,y_6} \quad \max_{i,k=1,\ldots,5} \left| f(s_i,t_k) - \sum_{r=1}^{6} y_r v_r(s_i,t_k) \right| \qquad (13)$$

where

$$s_i = \frac{i-1}{4}, \quad i = 1,\ldots,5$$

$$t_k = \frac{k-1}{4}, \quad k = 1,\ldots,5.$$

The problem (13) is then reformulated as a linear program, as described in §6. Then we get the task

Minimize y_7

subject to the linear constraints

$$\sum_{r=1}^{6} v_r(s_i,t_k)y_r + y_7 \geq f(s_i,t_k) \quad (i,k = 1,\ldots,5), \tag{14}$$

$$-\sum_{r=1}^{6} v_r(s_i,t_k)y_r + y_7 \geq -f(s_i,t_k) \quad (i,k = 1,\ldots,5). \tag{15}$$

This is a linear program with 7 variables and 50 constraints. It was solved with the simplex algorithm which is described in §12-14[*]. The following solution emerged:

$$
\begin{aligned}
y_1 &= 1.0358267 \\
y_2 &= -0.38764207 \\
y_3 &= -0.95174831 \\
y_4 &= -0.12398722 \\
y_5 &= 0.43169480 \\
y_6 &= 0.13390288 \\
y_7 &= 0.025910991.
\end{aligned}
\tag{16}
$$

The optimal solution is displayed in Fig. 17.3. Those vectors which appear in the optimal basis matrix are marked with a ⊕ or a ⊖ indicating the co-ordinates s_i,t_k of the corresponding mass-point. Here, ⊕ means that the basis vector gives equality in (14). Thus the error function

$$f - \sum_{r=1}^{n} y_r v_r$$

assumes a *maximum* value there with respect to the point set $\{(s_i,t_k)/i,k = 1,\ldots,5\}$. A ⊖ sign means that the basis vector gives equality in (15) and

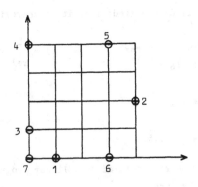

Fig. 17.3

---
[*]We thank Mr. Gerd Schuhfuss for solving this program, using a desk-calculator, an "HP 9825A".

hence the error function assumes a *minimum* value with respect to the same
set. The masses x_i of the points indicated in Fig. 17.3 are as follows:

$$
\begin{aligned}
x_1 &= 0.28571429 \\
x_2 &= 0.14285714 \\
x_3 &= -0.19047619 \\
x_4 &= 0.071428571 \\
x_5 &= -0.095238095 \\
x_6 &= -0.19047619 \\
x_7 &= -0.023809524.
\end{aligned}
\tag{17}
$$

The number q of mass-points which should result after Phase ii) of the
three-phase algorithm would in this case depend critically on the choice
of the tolerance ε. In particular, should the three points in the lower
left part of the figure be combined into one or more than one points?
To settle this question one could of course repeat Phase i); i.e. linear
programming with a finer grid. We take an alternative route. We consider
again the error function:

$$
g = f - \sum_{r=1}^{n} y_r v_r
$$

with $y_1, \ldots, y_6$ given by (16). Thus g satisfies $|g(s)| \le y_7$ on the
grid. We now determine the extremal points of g on the unit square
by means of a Newton-Raphson scheme and using appropriate grid-points as
starting approximations. Six local extrema were found. Their positions
are listed in the table below and marked in Fig. 17.4. There, θ means a
local minimum and $\oplus$ a local maximum.

Coordinates of local extrema

	0.288		4	0.000
1	0.000			1.000
	1.000		5	0.725
2	0.590			1.000
	0.000		6	0.867
3	0.118			0.000

$$\tag{18}$$

We now proceed to the construction of the nonlinear system based on the
assumption that the error curve corresponding to the solution of the continuous problem has its extremal points distributed in the same way as the
function g. (If this assumption should turn out to be wrong, then Phase

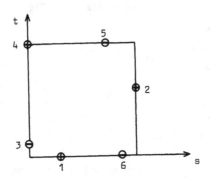

Fig. 17.4

iii) will fail and we must return to Phase i) which will be repeated with a refined grid.) Thus we put $q = 6$ and assume that all extremal points $(s_1,t_1),\ldots,(s_6,t_6)$ (numbering as in Fig. 17.4) lie on the boundary of the square. In particular we have $s_4 = 0$, $t_4 = 1$. The following 18 unknowns remain to be determined:

$$y_1,\ldots,y_6,y_7, \ x_1,\ldots,x_6, \ s_1,t_2,t_3,s_5,s_6. \tag{19}$$

From (8), (9), and (10) we get 13 equations. We note that the sign of the masses should be chosen as follows (see Fig. 17.4):

$$\text{sgn } x_1 = \text{sgn } x_2 = \text{sgn } x_4 = 1$$
$$\text{sgn } x_3 = \text{sgn } x_5 = \text{sgn } x_6 = -1.$$

Hence the 6 equations from (18) are completely determined in a simple form and (10) now becomes

$$x_1 + x_2 - x_3 + x_4 - x_5 - x_6 = 1.$$

The "missing" 5 equations are now generated from (11) since certain partial derivatives must vanish, e.g.

$$\frac{\partial}{\partial s} \left\{ f - \sum_{r=1}^{6} y_r v_r \right\}(s_1,0) = 0$$

$$\frac{\partial}{\partial t} \left\{ f - \sum_{r=1}^{6} y_r v_r \right\}(1,t_2) = 0.$$

Thus we have got 18 equations for the determination of the 18 unknowns of (19). An approximate solution was found as follows: s_1,t_2,t_3,s_5,s_6

were taken from (18) $y_1,\ldots,y_7$ from (16) and x_1,x_2,x_4,x_5,x_6 from (17), x_3 was calculated by combining the two masses numbered 3 and 7 of the solution of the discretized problem. Thus these two masses were added, giving $x_3 = -0.21428571$.

After four iterations the Newton-Raphson method delivered the results

$$y_1 = 0.98576860 \qquad y_4 = -0.14461987$$
$$y_2 = -0.34796776 \qquad y_5 = 0.42457304$$
$$y_3 = -0.90271418 \qquad y_6 = 0.11293036$$

and the maximal deviation was

$$y_7 = 0.027274796.$$

This agrees with the results reported by Andreasson and Watson. The solution of the dual problem is given in the table below:

Mass-point numbers	Coordinates s_i, t_i	Masses x_i
1	0.27210827 0.00000000	0.25885317
2	1.00000000 0.62068986	0.15098041
3	0.00000000 0.21767815	-0.20873218
4	0.0 1.0	0.090166417
5	0.67690452 1.00000000	-0.13844199
6	0.83562113 0.00000000	-0.15282583

(20) Remark. The procedure described above is applicable to many variants of the uniform approximation problem, e.g. when y must satisfy finitely or infinitely many linear constraints besides those specified in (PA). One example of such problems is one-sided approximation in the uniform norm. Further examples are to be found in Chapter IX.

We note that in many approximation problems

$$q = n+1$$

holds in (8), (9), and (10). (An important class of such problems is the case when $v_1,\ldots,v_n$ form a Chebyshev system on S. These problems will be treated in Chapter VIII.) If $q = n+1$, then $y_1,\ldots,y_{n+1}$ and $t_1,\ldots,t_{n+1}$ may be determined from (8) and (11) without also calculating

$x_1, \ldots, x_{n+1}$ from (9) and (10).

The mathematical properties of the system arising from (8) - (11) have been investigated in Hettich (1976), where nonlinear approximation is treated as well.

Chapter VIII

Approximation Problems by Chebyshev Systems

This chapter will be devoted to the study of the problem pairs (P) - (D) and (PA) - (DA) in a special but important case, namely when the moment generating functions $a_1, \ldots, a_n$ form a so-called Chebyshev system. The most well-known instance of such a system is $a_r(s) = s^{r-1}$, $r = 1, \ldots, n$, on a closed and bounded real interval. In all the linear optimization problems to be treated in this chapter, the structure of the nonlinear system can be determined from the outset, which simplifies the numerical treatment considerably in comparison to a direct application of the three-phase algorithm.

In the first section we shall present some general properties of Chebyshev systems. The reader will recognize many results from the theory of polynomials in one variable. The next section will be devoted to Problem (P) and the connection between one-sided approximation and certain generalized quadrature rules of the Gaussian type. In the last section we shall treat numerical calculation of the best approximations in the uniform norm.

§18. GENERAL PROPERTIES OF CHEBYSHEV SYSTEMS

(1) Let the functions $u_1, \ldots, u_n$ be continuous on the closed and bounded real interval $[\alpha, \beta]$. $u_1, \ldots, u_n$ will be called a *Chebyshev* system over $[\alpha, \beta]$, if the determinant

$$U(t_1,\ldots,t_n) = \begin{vmatrix} u_1(t_1) & \cdots & u_1(t_n) \\ \cdot & & \cdot \\ \cdot & & \cdot \\ \cdot & & \cdot \\ u_n(t_1) & \cdots & u_n(t_n) \end{vmatrix} \tag{2}$$

satisfies the relation

$$U(t_1,\ldots,t_n) > 0 \quad \text{if} \quad \alpha \le t_1 < t_2 < \ldots < t_n \le \beta. \tag{3}$$

Remark. The monominals $u_r(t) = t^{r-1}$, $r = 1,\ldots,n$ form a Chebyshev system over any real interval. See (3) of §7. From a numerical point of view it is often more advantageous to work with orthogonal polynomials instead of monomials. We note that if $a_r(t)$ is a polynomial of degree $r-1$ then we can determine constants d_r, $d_r = +1$ or $d_r = -1$, such that $u_r = d_r a_r$, $r = 1,\ldots,n$, is a Chebyshev system. The particular case $u_r = T_{r-1}$ (see (20) of §7) occurs often in computational practice. We also give the following example: Let λ_r be real numbers such that $0 \le \lambda_1 < \lambda_2 < \ldots < \lambda_n$. Put $u_r(t) = e^{\lambda_r t}$. Then $u_1,\ldots,u_n$ is a Chebyshev system over any real interval. The reader is referred to Karlin and Studden (1966) for further examples.

We will now show that many interesting results may be derived from the definiing relation (3).

(4) **Theorem.** Let $u_1,\ldots,u_n$ form a Chebyshev system over the closed and bounded interval $[\alpha,\beta]$. Let $w \in R^n$ be a fixed vector and let t_i, $i = 1,\ldots,n$ be distinct points. Then there is a unique vector $y \in R^n$ satisfying

$$\sum_{r=1}^{n} y_r u_r(t_i) = w_i, \quad i = 1,\ldots,n. \tag{5}$$

Proof: (5) is a linear system with n equations and the same number of unknowns. We may assume that the equations are reordered such that $t_1 < t_2 < \ldots < t_n$. By (3) we conclude that the system (5) has a nonzero determinant and hence a unique solution y.

We have immediately

(6) **Corollary.** Let $u_1,\ldots,u_n$ be as in (4) and define the function Q by

$$Q = \sum_{r=1}^{n} y_r u_r \tag{7}$$

where $y_1, \ldots, y_n$ are real numbers. Then Q has less than n zeros if Q is not identically zero.

Proof: Assume that Q is not identically zero but vanishes at $t_1, \ldots, t_n$. Putting $w_i = 0$, $i = 1, \ldots, n$ in (5) we get the unique solution $y_r = 0$, $r = 1, \ldots, n$, establishing the contradiction sought.

If we put $u_r(t) = t^{r-1}$, $r = 1, \ldots, n$, then Q becomes a polynomial of degree less than n. Thus (6) is a generalization of the familiar statement that a polynomial of degree less than n also has less than n zeros. In the same way, (4) generalizes the theorem that there is a unique polynomial of degree less than n which interpolates n given points.

(8) In order to discuss the problems (P) - (D) we also need to introduce zeros of multiplicity 2. We shall see that some well-known results on polynomials can easily be extended to linear combinations of functions which form a Chebyshev system. For this purpose we introduce the determinants

$$\tilde{U} \begin{pmatrix} j_1 & \cdots & j_n \\ t_1 & \cdots & t_n \end{pmatrix} \qquad\qquad (9)$$

where the symbols are defined by rules i), ii) and iii) below and the value is evaluated according to rules iv) and v).

i) $\alpha \le t_1 < t_2 < \ldots < t_n \le \beta$;

ii) j_i are integers and we have always $j_i = 1$ or $j_i = 2$;

iii) $j_1 = 1$; $j_i = 2$ is possible only if $j_{i-1} = 1$.

iv) If $j_1 = j_2 = \ldots = j_n = 1$ then

$$\tilde{U} \begin{pmatrix} j_1 & \cdots & j_n \\ t_1, & \ldots, & t_n \end{pmatrix} = U(t_1, \ldots, t_n)$$

v) If there is a $j_i = 2$ then we proceed as follows. First we assign to the determinant (9) the value given in Rule iv) above. Next we change all columns i with $j_i = 2$ so that the elements $u_r(t_i)$ are replaced by the divided difference $u_r[t_{i-1}, t_i]$ for $r = 1, \ldots, n$.

(10) Example.

$$\tilde{U} \begin{pmatrix} 1 & 1 & 2 \\ t_1 & t_2 & t_3 \end{pmatrix} = \begin{vmatrix} u_1(t_1) & u_1(t_2) & u_1[t_2, t_3] \\ u_2(t_1) & u_2(t_2) & u_2[t_2, t_3] \\ u_3(t_1) & u_3(t_2) & u_3[t_2, t_3] \end{vmatrix} .$$

We note that the determinants (2) and (9) generally have different numerical values (for the same points $t_1, \ldots, t_n$) if there is a $j_i = 2$. But (2) is positive if and only if (9) is positive.

(9) may also be *defined* if two arguments coincide. If the functions u_r are differentiable (which we assume from now on) then we define

$$u_r[t_i, t_i] = \lim_{t \to t_i} u_r[t, t_i] = u_r'(t_i).$$

We next introduce the determinant

$$U'(t_1, \ldots, t_n) \tag{11}$$

whose value is given by the rules a) - d) below.

a) $\alpha \leq t_1 \leq t_2 \leq \cdots \leq t_n \leq \beta$;

b) if $t_i = t_{i+1}$, then $t_i = \alpha$ or $t_i > t_{i-1}$
    and $t_{i+1} = \beta$ or $t_{i+1} < t_{i+2}$;

c) if all t_i are distinct, then $U'(t_1, \ldots, t_n) = U(t_1, \ldots, t_n)$;

d) if two arguments t_i coincide then we put

$$U'(t_1, \ldots, t_n) = \tilde{U}\begin{pmatrix} j_1 & \cdots & j_n \\ t_1, & \ldots, & t_n \end{pmatrix}$$

where

$$j_i = \begin{cases} 2 & \text{if } t_i = t_{i-1} \\ 1 & \text{if } t_i > t_{i-1}. \end{cases}$$

(12) Example.

$$U'(t_1, t_2, t_2) = \tilde{U}\begin{pmatrix} 1 & 1 & 2 \\ t_1 & t_2 & t_2 \end{pmatrix} = \begin{vmatrix} u_1(t_1) & u_1(t_2) & u_1'(t_2) \\ u_2(t_1) & u_2(t_2) & u_2'(t_2) \\ u_3(t_1) & u_3(t_2) & u_3'(t_2) \end{vmatrix}.$$

(13) The functions $u_1, \ldots, u_n$ are said to form an *extended Chebyshev system of order two* over $[\alpha, \beta]$ if $u_1, \ldots, u_n$ are continuously differentiable on $[\alpha, \beta]$ and all determinants $U'(t_1, \ldots, t_n) > 0$ for $\alpha \leq t_1 \leq \cdots \leq t_n \leq \beta$.

(14) A function f which is continuously differentiable on $[\alpha, \beta]$ is said to have a zero of *multiplicity 2* (also called a *double zero*) at $t \in [\alpha, \beta]$, if $f(t) = 0$ and $f'(t) = 0$. We can now extend the Corollary (6) and state:

(15) Theorem. Let $u_1, \ldots, u_n$ form an extended Chebyshev system of order two over $[\alpha, \beta]$. Let the linear combination Q be given by (7).

Assume also that Q is not identically zero. Then Q has less than n
zeros in $[\alpha,\beta]$, counted with multiplicity.

Proof: The proof is analogous to that of (6). We note that if Q
has a double zero at $\bar{t}$, then the coefficient vector y of Q must
satisfy the two equations

$$Q(\bar{t}) = 0, \quad Q'(\bar{t}) = 0.$$

Thus if we assume that Q has n zeros counted with multiplicity we get
a linear system of equations whose right-hand side is zero and whose coefficient matrix has a determinant of the type (11). Hence the conclusion
follows.

Remark. The interpolation statement (5) may be extended to the con-
fluent case, i.e. when pairs of the points t_i appearing in (15) are
allowed to coincide. One could also introduce extended Chebyshev systems
of order higher than 2 and establish the corresponding results on interpolation and maximum number of zeros.

Some results which will be needed in the sequel are given in the
exercises below.

(16) Exercise. Let $u_1,\ldots,u_{n+1}$ as well as $u_1,\ldots,u_n$ be extended
Chebyshev systems of the second order over $[\alpha,\beta]$. Show that

$$U'(t_1,\ldots,t_n,t) = c_n(u_{n+1}(t) - Q(t)) \tag{17}$$

where $c_n > 0$ is independent of t and Q is a linear combination of
$u_1,\ldots,u_n$ such that

i) $Q(t_i) = u_{n+1}(t_i)$, $i = 1,\ldots,n$;

ii) if $t_i = t_{i+1}$, then $Q'(t_i) = u'_{n+1}(t_i)$.

Hint: Expand the determinant $U'(t_1,\ldots,t_n,t)$ by the last column.

(18) Exercise. Show that if we take $u_r(t) = t^{r-1}$ and require
that u_{n+1} has n continuous derivatives and also satisfies $u_{n+1}^{(n)}(t) > 0$,
$t \in [\alpha,\beta]$, then the functions $u_1,\ldots,u_{n+1}$ satisfy the assumptions of
Exercise (16). Hint: Use Rolle's theorem to show that no function of
the form

$$u_{n+1}(t) - \sum_{r=1}^{n} y_r u_r(t)$$

can have n+1 zeros in $[\alpha,\beta]$.

(19) _Exercise_. Use the notation and assumptions of (16). Let

$$R(t) = u_{n+1} - Q(t).$$

Show the following results:

$R(t) \geq 0$, $t \in [\alpha,\beta]$, in the two cases i) and ii) below:

i)   R has n/2 double zeros in (α,β) (n even);

ii) $R(\alpha) = 0$ and R has (n-1)/2 double zeros in (α,β) (n odd).

And $R(t) \leq 0$, $t \in [\alpha,\beta]$, in the two cases i) and ii) below:

i) $R(\alpha) = R(\beta) = 0$, R has (n/2)-1 double zeros in (α,β) (n
     even);

ii) $R(\beta) = 0$ and R has (n-1)/2 double zeros in (α,β) (n odd).

Hint: R does not change sign at a double zero.

(20) _Exercise_. Let $u_1,\ldots,u_{n+1}$ be an extended Chebyshev system of
order two over $[\alpha,\beta]$. Show that there is a linear combination

$$Q = u_{n+1} - \sum_{r=1}^{n} y_r u_r \tag{21}$$

which is strictly positive on $[\alpha,\beta]$. _Hint_: If Q has only double zeros
in (α,β) and possibly simple zeros at α and β then either $Q(t) \geq 0$,
$t \in [\alpha,\beta]$, or $-Q(t) \geq 0$, $t \in [\alpha,\beta]$. Put $2Q = Q_1 + Q_2$ where Q_1 and
Q_2 are nonnegative linear combinations of the type of (21). The cases
n odd and n even should be discussed separately.

§19. ONE-SIDED APPROXIMATION AND GENERALIZED QUADRATURE RULES OF THE GAUSSIAN TYPE

In this section we shall study Problem (P) - (D), which we defined in
§3 and §4. Here we study a special but important case. The index set S
is the closed and bounded interval $[\alpha,\beta]$ and $a_1,\ldots,a_n$ form an extended
Chebyshev system over $[\alpha,\beta]$. b will be assumed to be continuously differentiable on S. Instead of a_r we shall write u_r, as in §18. Sometimes we shall assume that the n+1 functions $u_1,\ldots,u_n$, b also form
an extended Chebyshev set of order two over S. Then we shall write u_{n+1}
instead of b. Besides the dual pair (P) - (D) we shall study the two
problems

(P_2) Maximize $\sum_{r=1}^{n} c_r y_r$ subject to $\sum_{r=1}^{n} u_r(s) y_r \leq b(s)$, $s \in S$,

and

(D_2) Minimize $\displaystyle\sum_{i=1}^{q} x_i b(s_i)$ subject to $\displaystyle\sum_{i=1}^{q} x_i u(s_i) = c,$

$$x_i \geq 0, \quad i = 1,\ldots,q.$$

(1) **Exercise.** Show that (P_2) - (D_2) are a dual pair of linear optimization problems.

(2) **Lemma.** Let $u_1,\ldots,u_n$ form an extended Chebyshev set over the closed and bounded interval S and let b be continuous there. Then (P) and (P_2) meet the Slater condition.

Proof: Using the result of (20) of §18 we establish that there is a vector $z \in R^n$ such that

$$Q(s) = \sum_{r=1}^{n} z_r u_r(s) > 0, \quad s \in S.$$

Put

$$d = \min_{s \in S} Q(s)$$

Next set

$$\lambda = d^{-1}(1 + \max_{s \in S} b(s))$$

and define the vector $y = \lambda z$. Then we get

$$\sum_{r=1}^{n} y_r u_r(s) = \lambda \sum_{r=1}^{n} z_r u_r(s) \geq \lambda d > \max_{s \in S} b(s).$$

Thus (P) meets the Slater condition. The proof of the analogous statement for (P_2) is carried out in a similar way.

(3) Hence we may use (12) of §10 to conclude that if (D) is feasible and the assumptions of Lemma (2) are met, then strong duality holds for (P) - (D) and (P_2) - (D_2) respectively. If we also require that c is in the interior of the moment cone M_n, we can show that (P) has a unique solution. (For a definition of this cone see §8.) For this purpose we shall give a simple characterization of interior and boundary points of the moment cone. Our argument parallels that of Karlin-Studden (1966), pp. 42-43.

(4) Let $T = \{t_1,\ldots,t_\ell\}$ be a subset of the interval $[\alpha,\beta]$. Then we shall denote by the *index of* T, ind (T), the integer

$$\text{ind } (T) = \sum_{i=1}^{\ell} \{\text{sign}(t_i-\alpha) + \text{sign}(\beta-t_i)\}.$$

Thus ind (T) must assume one of the three values $2\ell - 2$, $2\ell - 1$, 2ℓ.

(5) <u>Example</u>. Consider the function

$$R = u_{n+1} - Q$$

of (19) of §18. Let $Z = \{z_1,\ldots,z_q\}$ denote its zeros. Then we can select Q in such a manner that ind $(Z) = n$ and $R(s) \leq 0$, $s \in [\alpha,\beta]$, or ind $(Z) = n$ and $R(s) \geq 0$, $s \in [\alpha,\beta]$.

We discuss the special instance $S = [0,1]$, $u_r(s) = s^{r-1}$, $r = 1,\ldots,n$, and $u_{n+1}(s) = e^s$. By (18) of §18, R has at most n zeros in [0,1], counted with multiplicity. We want to construct R explicitly for $n = 3$ and $n = 4$ and indicate the zeros of R. The zeros in (0,1) will be double. They are denoted z_i, $i = 1,\ldots$. Thus

$n = 4$	$Z = \{z_1,z_2\}$	ind$(Z) = 4$	$R(s) \geq 0$, $s \in [0,1]$;
$n = 4$	$Z = \{0,z_3,1\}$	ind$(Z) = 4$	$R(s) \leq 0$, $s \in [0,1]$;
$n = 3$	$Z = \{0,z_4\}$	ind$(Z) = 3$	$R(s) \geq 0$, $s \in [0,1]$;
$n = 3$	$Z = \{z_5,1\}$	ind$(Z) = 3$	$R(s) \leq 0$, $s \in [0,1]$.

(6) Let $c \in M_n$ be a given vector. If

$$c_r = \sum_{i=1}^{\ell} x_i u_r(s_i), \quad r = 1,\ldots,n, \tag{7}$$

$$x_i > 0, \quad i = 1,\ldots,q,$$

then we say that c has a *representation* involving the points $s_1,\ldots,s_q$. We define now the *index of c* as ind(T) where T is the subset with smallest index satisfying (7).

(8) <u>Example</u>. $S = [0,1]$, $u_r(s) = s^{r-1}$, $r = 1,\ldots,4$. $c = (1,1/2, 1/3,1/4)^T$ has the two representations

$$c = \frac{1}{6} u(0) + \frac{2}{3} u(1/2) + \frac{1}{6} u(1) \tag{9}$$

and

$$c = \frac{1}{2} u(\frac{1}{2} - \frac{1}{\sqrt{12}}) + \frac{1}{2} u(\frac{1}{2} + \frac{1}{\sqrt{12}}). \tag{10}$$

The index of the subsets appearing in (9) and (10) is 4.

(11) <u>Exercise</u>. Show that for c from (8) we do have ind (c) = 4. <u>Hint</u>: all subsets with index 3 must be of one of the two forms $\{0,t\}$ or $\{t,1\}$, where $t \in (0,1)$.

(12) <u>Lemma</u>. Let $u_1,\ldots,u_n$ be an extended Chebyshev system of order two over S. A point $c \in M_n$ is a boundary point of M_n if and

only if $ind(c) < n$. Every boundary point admits a unique representation (7).

   Proof: Let $c^0 \in Bd\,(M_n)$. Then there is a supporting hyperplane to M_n at c^0, passing through 0 since M_n is a convex cone. Thus we may find real constants β_r such that

$$\sum_{r=1}^{n} \beta_r^2 > 0$$

satisfying

$$\sum_{r=1}^{n} \beta_r c_r^0 = 0, \quad \sum_{r=1}^{n} \beta_r c_r \geq 0, \quad c \in M_n. \tag{13}$$

Now put

$$Q = \sum_{r=1}^{n} \beta_r u_r.$$

By (13) we get

$$Q(t) \geq 0, \quad t \in [\alpha, \beta].$$

Since $c^0 \in M_n$, c^0 must have a representation (7). We get

$$\sum_{r=1}^{n} c_r^0 \beta_r = \sum_{i=1}^{\ell} x_i \sum_{r=1}^{n} \beta_r u_r(s_i) = \sum_{i=1}^{\ell} x_i Q(t_i) = 0.$$

Thus $Q(t_i) = 0$, $i = 1,\ldots,$. Theorem (15) of §18 can be reformulated to the statement that the set of zeros of Q has an index $< n$. Hence $ind(c_0) < n$. The numbers x_i in (7) are uniquely determined as long as $\ell \leq n$, since if $\ell < n$ we add $t_{\ell+1},\ldots,t_n$ to the sum in (7), where $t_{\ell+1},\ldots,t_n$ are selected such that $t_1,\ldots,t_n$ are distinct. We put $x_{\ell+1} = \ldots = x_n = 0$. We next consider (7) as a linear system of equations with $x_1,\ldots,x_n$ as unknowns. It has a unique solution since its determinant is positive.

   Assume conversely that a vector $c \in M_n$ has a representation (7) with index $< n$. We construct a nonnegative function Q whose coefficients define a supporting hyperplane at c. By (2) of §11, c must be a boundary point of M_n. This concludes the proof.

   (14) Theorem. Let $u_1,\ldots,u_n$ be an extended Chebyshev system of order 2 over S, and let b have a continuous derivative on S. If (D) is feasible then (D) and (D_2) have optimal solutions. If c is in the interior of the moment cone M_n then (P) and (P_2) have unique solutions.

Proof: Since the Slater condition is met in P and P_2, Problems (D) and (D_2) have optimal solutions. Let $\bar{\gamma}$ be the optimal value of (D). Then there are nonnegative reals x_i and elements $t_i \in S$ such that

$$\sum_{i=1}^{q} x_i u_r(t_i) = c_r, \quad r = 1,\ldots,n,$$

$$x_i \geq 0, \quad i = 1,\ldots,q.$$

If $c \in \overset{o}{M}_n$, then (P) has an optimal solution y by Theorem (7) of §11. Put

$$Q = \sum_{r=1}^{n} y_r u_r.$$

Due to complementary slackness we have

$$Q(t_i) = b(t_i), \quad i = 1,\ldots,q. \tag{15}$$

Since $Q(s) \geq b(s)$, $s \in S$, we must also require

$$Q'(t_i) = b'(t_i) \quad \text{if} \quad \alpha < t_i < \beta. \tag{16}$$

Combining (15) and (16) we get a linear system of equations with n unknowns and a number of equations amounting to ind $(t_1,\ldots,t_q)$. Since $c \in \overset{o}{M}_n$ we conclude from Lemma (12) that ind $(c) \geq n$. Thus (15) and (16) uniquely determine the optimal solution y.

(17) **Example.** If c is at the boundary of M_n, then (P) may not have any solution or there might be many solutions. Consider the problem

Minimize y_1 subject to $y_1 + y_2 s + y_3 s^2 \geq b(s)$, $s \in [-1,1]$

where b is a function continuously differentiable on $[-1,1]$.

The dual of this problem reads

Maximize $\sum_{i=1}^{q} x_i b(t_i)$ subject to

$$\sum_{i=1}^{q} x_i = 1, \tag{18}$$

$$\sum_{i=1}^{q} x_i t_i = 0, \tag{19}$$

$$\sum_{i=1}^{q} x_i t_i^2 = 0, \quad x_i \geq 0, \ i = 1,\ldots,q,$$

$$-1 \leq t_i \leq 1, \quad i = 1,\ldots,q. \tag{20}$$

Combining (18) and (20) we find that we must take $q = 1$, $t_1 = 0$, $x_1 = 1$. Thus ind $(1,0,0)^T = 2$ for this problem. Let now $y \in R^3$ be given and put $Q(s) = y_1 + y_2 s + y_3 s^2$. y is optimal if and only if

$Q(0) = b(0)$,

$Q(s) \geq b(s)$, $\quad s \in [-1,1]$.

Thus we must also have $Q'(0) = b'(0)$.

Hence a solution y must satisfy

$y_1 = b(0)$ $\quad y_2 = b'(0)$ $\quad$ and

$y_3 s^2 \geq b(s) - b(0) - sb'(0)$, $\quad -1 \leq s \leq 1$. $\hspace{2cm}$ (21)

y_3 is generally not determined uniquely by (21). For $f(s) = \exp(s)$ we get the condition $y_3 \geq e-2 \approx 0.718$.

In the case $f(s) = |s|^{3/2}$, (21) gives the relation

$y_3 s^2 \geq |s|^{3/2}$, $\quad -1 \leq s \leq 1$

which cannot be satisfied for any y. Thus (P) has no solution in this case.

The conditions of Theorem (14) do not, however, guarantee the uniqueness of solutions to (D). This is illustrated by

(22) **Example**.

(P) $\quad$ Minimize $y_1 + \frac{1}{2} y_2$ subject to $y_1 + y_2 s \geq 1 + s \cos 6\pi s$, $\quad 0 \leq s \leq 1$.

The dual of this problem reads

(D) $\quad$ Maximize $\sum_{i=1}^{q} x_i (1 + s_i \cos 6\pi s_i)$ $\quad$ subject to

$\sum_{i=1}^{q} x_i = 1$,

$\sum_{i=1}^{q} x_i s_i = 1/2$,

$x_i \geq 0$, $\quad s_i \in [0,1]$, $\quad i = 1,\ldots,q$.

We can take $q = 1$, $x_1 = 1$, $s_1 = 1/2$. Thus ind $(1,\frac{1}{2})^T = 2$ in this problem, i.e. $(1,\frac{1}{2})^T \in \overset{\circ}{M}_2$. Taking $y_1 = 3$, $y_2 = 0$ we find that the Slater condition is met. By (14), (P) has a unique solution. We note that

$1 + s \cos 6\pi s \leq 1 + s$,

with equality at $s = 0$, $1/3$, $2/3$, 1. Hence an optimal solution to (D) is defined by the conditions

$$\sum_{i=1}^{4} x_i = 1,$$

$$\sum_{i=1}^{4} x_i \frac{i-1}{3} = \frac{1}{2}, \quad x_i \geq 0, \quad i = 1,\ldots,4.$$

These conditions do not determine $x_1,\ldots,x_4$ uniquely.

(23) <u>Theorem</u>. Let $u_1,\ldots,u_n$ as well as $u_1,\ldots,u_{n+1}$ be extended Chebyshev systems of order two over S. If $c \in M_n$, then (D) and (D_2) have unique solutions. (P) and (P_2) have solutions which are uniquely determined if $c \in \overset{\circ}{M}_n$.

<u>Proof</u>: The statements about the solutions of (P) and (P_2) for $c \in \overset{\circ}{M}_n$ are a direct consequence of Theorem (14). We now treat the case $c \in$ bd M_n and study the solutions of (D). Let c have the representation

$$c_r = \sum_{i=1}^{q} x_i u_r(t_i), \quad r = 1,\ldots,n. \tag{24}$$

If $c \in$ bd M_n then ind (c) < n and (24) is uniquely determined by c. Then there is only one subset $\{t_1,\ldots,t_q\}$ such that the constraints of (D) are met, so (D) has trivially a unique optimal solution. We next show that (P) has a solution y. Points $t_{q+1},\ldots,t_\ell$ are selected in such a manner that

$$\text{ind } \{t_1,\ldots,t_\ell\} = n$$

and this set contains the endpoint β. Next, y is determined from the equations

$$y^T u(t_i) = u_{n+1}(t_i), \quad i = 1,\ldots,\ell,$$

$$y^T u'(t_i) = u'_{n+1}(t_i), \quad t_i \pm (\alpha,\beta), \quad i = 1,\ldots,\ell,$$

where $u(t_i) = (u_1(t_i),\ldots,u_n(t_i))^T$.

As shown in (19) of §18, y meets the constraints of (P). The construction of a solution to (P_2) proceeds in a similar manner. We need to show that (D) has a unique solution if $c \in \overset{\circ}{M}_n$. Let $\bar{\lambda}$ be the optimal value of (D), $\underline{\lambda}$ the optimal value of (D_2). Then $\underline{\lambda} \leq \bar{\lambda}$. Since M_{n+1} is closed, the optimal values are attained. Also,

$$(c_1, \ldots, c_n, \bar{\lambda}) \in \text{bd } M_{n+1}.$$

Hence it has a unique representation given by

$$\sum_{i=1}^{\bar{q}} \bar{x}_i u(\bar{t}_i) = c, \qquad \sum_{i=1}^{\bar{q}} \bar{x}_i u_{n+1}(\bar{t}_i) = \bar{\lambda}, \tag{25}$$

and we have $\text{ind } (\bar{t}_1, \ldots, \bar{t}_{\bar{q}}) \leq n$. (D_2) is treated in the same way. Thus we have concluded the proof.

(26) **Remark.** If $c \in \overset{\circ}{M}_n$ then $\text{ind } (c) \geq n$. Combining this knowledge with (25) we get

$$\text{ind } \{\bar{t}_1, \ldots, \bar{t}_{\bar{q}}\} = n.$$

If we discuss (D_2) in the same way we shall find a representation

$$\sum_{i=1}^{q} x_i u(t_i) = c, \qquad \sum_{i=1}^{q} x_i u_{n+1}(t_i) = \underline{\lambda}, \tag{27}$$

where

$$\text{ind } \{t_1, \ldots, t_q\} = n.$$

Since (P) and (P_2) have unique solutions we must have $\underline{\lambda} < \bar{\lambda}$. Thus (25), (27) defines two different representations of c. We note also that if $c \in \overset{\circ}{M}_n$ then (P) has a unique optimal solution y. Put

$$Q = \sum_{r=1}^{n} y_r u_r.$$

Then we must have

$$Q(\bar{t}_i) = u_{n+1}(\bar{t}_i), \qquad i = 1, \ldots, \bar{q},$$

$$Q(t) \geq u_{n+1}(t).$$

Therefore the right endpoint β must be in the subset $\{\bar{t}_1, \ldots, \bar{t}_q\}$. (See (19) of §18.) Arguing in the same way we find that the set $\{t_1, \ldots, t_q\}$ is also uniquely determined and does not contain the endpoint β. Thus if $c \in \overset{\circ}{M}_n$ then c has two different representations associated with subsets of index n.

(28) **Generalized quadrature rules of the Gaussian type.** Let again $u_1, \ldots, u_n$ form an extended Chebyshev system of order two over $[\alpha, \beta]$ and let w be a continuous nonnegative function over the same interval. For functions f which are continuously differentiable over $[\alpha, \beta]$ we define

$$I(f) = \int_{\alpha}^{\beta} f(s)w(s)ds.$$

We want to construct mechanical quadrature rules of the form

$$I(f) \approx \sum_{i=1}^{q} x_i f(s_i) \tag{29}$$

where $\alpha \le s_1 < s_2 < \ldots < s_q \le \beta$.

We want (29) to give exact results for $f = u_r$, $r = 1,\ldots,n$. Putting

$$c_r = I(u_r) = \int_{\alpha}^{\beta} u_r(s)w(s)ds, \quad r = 1,\ldots,n, \tag{30}$$

we find that the weights x_i and the abscissas s_i must meet the condition

$$\sum_{i=1}^{q} x_i u_r(s_i) = c_r, \quad r = 1,\ldots,n. \tag{31}$$

If we put $q = n$ in (31) and select s_i arbitrarily we may consider (31) as a linear system with $x_1,\ldots,x_n$ as unknowns. Since $u_1,\ldots,u_n$ form a Chebyshev system, the determinant of this system is positive and hence a unique solution exists.

We now show that there are exactly two rules (31) such that $x_i \ge 0$, $i = 1,\ldots,q$, and ind $(s_1,\ldots,s_q) = n$. These rules are called generalized rules of the Gaussian type. To establish this we need only show that $(c_1,\ldots,c_n)^T \in M_n$, since then we can apply the argument of (26). For $N = 2,3,\ldots$ we define for $r = 1,\ldots,n$ the functions $\tilde{u}_{rN}$ according to

$$\tilde{u}_{rN}(s) = \begin{cases} u_r(\alpha), & s = \alpha \\[2mm] u_r\{\dfrac{(N-i)\alpha+i\beta}{N}\}, & \alpha + \dfrac{i-1}{N}(\beta-\alpha) < s \le \alpha + \dfrac{i(\beta-\alpha)}{N}. \end{cases}$$

We find that

$$\begin{aligned} &\lim_{N\to\infty} \tilde{u}_{rN}(s) = u_r(s), \quad r = 1,\ldots,n, \\[2mm] &\lim_{N\to\infty} \int_{\alpha}^{\beta} \tilde{u}_{rN}(s)w(s)ds = c_r, \quad r = 1,\ldots,n. \end{aligned} \tag{32}$$

Put

$$c_r^N = \int_{\alpha}^{\beta} \tilde{u}_{rN}(s)w(s)ds.$$

We find that

$$c_r^N = \sum_{i=1}^{N} \xi_i u_r \{\frac{(N-i)\alpha+i\beta}{N}\} ,$$

where

ξ_i is the integral of w over the interval

$[\alpha + (i-1)(\beta-\alpha)/N, \ \alpha + i(\beta-\alpha)/N]$.

Thus $c^N = (c_1^N,\ldots,c_2^N)^T \in M_n$, $N = 1,\ldots$. Since M_n is closed, $c \in M_n$ due to (32).

(33) <u>One-sided approximation</u>. Let $u_1,\ldots,u_n [\alpha,\beta]$ be as in (28). We discuss now the problem of approximating the continuously differentiable function f from above by the linear combination

$$Q = y^T u$$

in such a manner that

$$\int_\alpha^\beta |Q(s) - f(s)| w(s) ds \qquad\qquad (34)$$

is minimized when $Q(s) \geq f(s)$, $s \in S$. Here w is a fixed function, continuous on $[\alpha,\beta]$. Since $Q(s) \geq f(s)$, $|Q(s) - f(s)| = Q(s) - f(s)$ and (34) becomes

$$\int_\alpha^\beta |Q(s) - f(s)| w(s) ds = c^T y - \int_\alpha^\beta f(s) w(s) ds, \qquad (35)$$

where c is given by (30). Since the integral on the right hand side of (35) is independent of y, our goal is to render the scalar product a minimum subject to the constraint $Q(s) \geq f(s)$, $s \in S$. We recognize an instance of (P). We note that q, $\{s_1,\ldots,s_q\}$, $x_1,\ldots,x_q$ is feasible for the dual problem (D) if and only if $s_1,\ldots,s_q$ and $x_1,\ldots,x_q$ are the abcissas and weights of a quadrature rule (with nonnegative weights) which is exact for $u_1,\ldots,u_n$. By complementary slackness the optimal Q must satisfy the equations

$$Q(s_i) = f(s_i), \quad i = 1,\ldots,q, \qquad\qquad (36)$$

$$(s_i-\alpha)(\beta-s_i)Q'(s_i) = f'(s_i) = 0, \quad i = 1,\ldots,q. \qquad (37)$$

If the n+1 functions $u_1,\ldots,u_n,f$ form an extended Chebyshev system of order two then the optimal solutions of (D) and (D_2) define generalized rules of the Gaussian type. See (26).

(38) Example. We want to find the best polynomial approximations
from above and below to the function e^t on $[0,1]$ for $w(t) = 1$. We
treat the cases $n = 3,4$ and $u_r(s) = s^{r-1}$. Thus $c = (1,1/2,1/3,1/4)^T$.
For $n = 4$ there are two (generalized) rules of Gaussian type which can
be found from (9) and (10). Thus the best approximation to e^t from
above is found by solving (36), (37) with $q = 3$, $s_1 = 0$, $s_2 = 1/2$, $s_3 = 1$,
$\alpha = 0$, $\beta = 1$. See also (5). For $n = 3$ the two generalized Gaussian
rules have the abscissas 0, $2/3$ and $1/3$, 1 respectively.

(39) Calculation of generalized quadrature rules of the Gaussian
type. Such rules can be determined by solving (P), (D) for $a_r = u_r$ and
$b = u_{n+1}$, where $u_1,\ldots,u_{n+1}$ and $u_1,\ldots,u_n$ are required to form extended Chebyshev systems of order two over $[\alpha,\beta]$. The three-phase algorithm is simplified considerably since q is known from the outset.
It is also known that $t_q = \beta$ must occur in the representation sought
for c. When n is even we also have $t_1 = \alpha$. Thus the structure of
the nonlinear system treated in Phase 3 is known from the outset and we
know for certain whether a "correct" system has been constructed after
carrying out Phases 1 and 2. We observe that $s_1,\ldots,s_q$ and $x_1,\ldots,x_q$
can be found from the nonlinear system (4) of §16 which in this case has
n equations and n unknowns. If one wants to solve (P) instead, y
can afterwards be found from the linear system resulting from combining
(3) and (5) of §16. For the important case $u_r(s) = s^{r-1}$ special algorithms have been developed.

§20. COMPUTING THE BEST APPROXIMATION IN THE UNIFORM NORM

In this section we shall treat the numerical solution of the dual
pair (PA) - (DA) when $v_1,\ldots,v_n$ form an extended Chebyshev system of
order two over an interval $[\alpha,\beta]$ and f is twice differentiable over
the same interval. Instead of v_r we shall write u_r, $r = 1,\ldots,n$. We
write (PA) and (DA) as follows (see §6):

(PA)   Minimize y_{n+1} subject to $\left| \sum_{r=1}^{n} y_r u_r(t) - f(t) \right| \le y_{n+1}$, $t \in [\alpha,\beta]$;

(DA)   Maximize $\sum_{i=1}^{q} x_i f(t_i)$ subject to

$$\sum_{i=1}^{q} x_i u_r(t_i) = 0, \quad r = 1,\ldots,n,$$

$$\sum_{i=1}^{q} |x_i| = 1.$$

In §7 we treated polynomial approximation; i.e. the case $u_r(t) = t^{r-1}$. We shall now show that many of the results obtained there may be easily extended to case of a general extended Chebyshev system of order two.

(1) __Lemma.__ Let $\alpha \leq t_1 < t_2 < \ldots < t_{n+1} \leq \beta$ be fixed real numbers and let $x_1, \ldots, x_{n+1}$ be a nontrivial solution of the homogeneous system of equations

$$\sum_{i=1}^{n+1} u_r(t_i) x_i = 0, \quad r = 1, \ldots, n. \tag{2}$$

Then

$$x_i x_{i+1} < 0, \quad i = 1, \ldots, n.$$

__Proof:__ Let i be a fixed integer such that $1 \leq i \leq n$. Let

$$P = \sum_{r=1}^{n} y_r u_r,$$

the linear combination which is uniquely determined by the conditions

$$P(t_j) = \begin{cases} 1, & j = i, \\ 0, & j = 1, \ldots, n+1, \quad j \neq i, \quad j \neq i+1. \end{cases} \tag{3}$$

The determinant of the system of equations (3) is positive by the definition of Chebyshev systems. The rest of the argument parallels the proof of Lemma (1) of §7.

(4) __Theorem.__ Let f be continuous on $[\alpha, \beta]$, $v_1, \ldots, v_n$ a Chebyshev system on the same interval and a linear combination P be given:

$$P = \sum_{r=1}^{n} y_r v_r.$$

Let further $\alpha \leq t_1 < t_2 < \ldots < t_{n+1} \leq \beta$ be $n+1$ points such that

$$\{f(t_i) - P(t_i)\} \cdot \{f(t_{i+1}) - P(t_{i+1})\} < 0, \quad i = 1, \ldots, n. \tag{5}$$

Then

$$\min_i |f(t_i) - P(t_i)| \leq \Delta_n \leq \max_{\alpha \leq t \leq \beta} |f(t) - P(t)|, \tag{6}$$

where

$$\Delta_n = \inf_{y \in R^n} \max_{\alpha \leq t \leq \beta} \left| f(t) - \sum_{r=1}^{n} y_r u_r \right|.$$

__Proof:__ The proof closely follows that of Theorem (5) of §7 if we replace t^{r-1} with u_r there.

(7) <u>Corollary</u>. Let

$$P = \sum_{r=1}^{n} y_r u_r$$

be a linear combination such that there are n+1 points $\alpha \leq t_1 < t_2 <$
$\ldots < t_{n+1} \leq \beta$ with the properties

$$|\delta_i| = |f(t_i) - P(t_i)| = \max_{\alpha \leq t < \beta} |f(t) - P(t)|, \quad i = 1,\ldots,n+1,$$

and

$$\delta_i \cdot \delta_{i+1} < 0, \quad i = 1,\ldots,n.$$

Then P is the linear combination of $u_1,\ldots,u_n$ which best approximates
f in the uniform norm.

(8) <u>Determination of a linear combination satisfying (5)</u>. Again let
$\alpha \leq t_1 < t_2 < \ldots < t_{n+1} \leq \beta$ be given. We seek a linear combination P
of $u_1,\ldots,u_n$ and a constant ε such that

$$P(t_i) = f(t_i) - \varepsilon(-1)^i, \quad i = 1,\ldots,n+1. \tag{10}$$

Putting

$$P = \sum_{r=1}^{n} y_r u_r, \tag{11}$$

we get the linear system of equations

$$\sum_{r=1}^{n} y_r u_r(t_i) + \varepsilon(-1)^i = f(t_i), \quad i = 1,\ldots,n+1. \tag{12}$$

There are n+1 unknowns, namely $y_1,\ldots,y_n$, and ε, and the same number
of equations. Expanding the determinant of coefficients by its last
column and using the defining property (2) of §18 of Chebyshev systems,
we ascertain that (12) has a unique solution.

For a general Chebyshev system, (12) may be solved numerically as
described in §14 but if $u_r(t) = t^{r-1}$ the method of (14) of §7 is faster.
If $u_r = T_{r-1}$, $r = 1,\ldots,n$ and

$$t_i = \frac{a+b}{2} + \frac{b-a}{2} \cos \frac{i-1}{n} \pi, \quad i = 1,\ldots,n+1,$$

then the orthogonality relations (34) of §7 can be used to solve (12)
efficiently.

By Theorem (4), $|\varepsilon|$ from (10) is a lower bound for the value of
(PA). We next describe how to improve upon this bound in a systematic
manner.

(13) <u>Lemma</u>. Let $u_1,\ldots,u_n$ form a Chebyshev system over $[\alpha,\beta]$ and let f be continuous on the same interval. Let $\alpha \le t_1 < t_2 < \ldots < t_{n+1} \le \beta$ be given and let $y \in R^n$ and ε be the solution of (12). Put

$$R(t) = f(t) - \sum_{r=1}^{n} y_r u_r(t).$$

Now let $\alpha \le \tau_1 < \tau_2 < \ldots < \tau_{n+1} \le \beta$ be such that

$$R(\tau_i) \cdot R(\tau_{i+1}) < 0, \quad i = 1,\ldots,n; \tag{14}$$

$$|R(\tau_i)| \ge |R(t_i)|, \quad i = 1,\ldots,n+1; \tag{15}$$

$$|R(\tau_i)| > |R(t_i)| \quad \text{for at least one } i. \tag{16}$$

We define $z \in R^n$ and $\bar{\varepsilon}$ through

$$\sum_{r=1}^{n} z_r u_r(\tau_i) + \bar{\varepsilon}(-1)^i = f(\tau_i), \quad i = 1,\ldots,n+1. \tag{17}$$

Then $|\bar{\varepsilon}| > |\varepsilon|$.

<u>Proof</u>: We determine $x_1,\ldots,x_{n+1}$ and $\xi_1,\ldots,\xi_{n+1}$ as the unique solutions of the equations

$$\sum_{i=1}^{n+1} x_i u_r(t_i) = 0, \quad r = 1,\ldots,n, \quad \sum_{i=1}^{n+1} (-1)^i x_i = 1, \tag{18}$$

$$\sum_{i=1}^{n+1} \xi_i u_r(\tau_i) = 0, \quad r = 1,\ldots,n, \quad \sum_{i=1}^{n+1} (-1)^i \xi_i = 1. \tag{19}$$

Since the matrix of coefficients of (18) is the transpose of that of (12) the former system has a unique solution. This is true of (19) by the same argument. From Lemma (1) we conclude that

$$x_i x_{i+1} < 0, \quad \xi_i \xi_{i+1} < 0, \quad i = 1,2,\ldots,n. \tag{20}$$

Multiplying (17) by ξ_i and summing over i we get

$$\sum_{i=1}^{n+1} \xi_i \sum_{r=1}^{n} z_r u_r(\tau_i) + \bar{\varepsilon} \sum_{i=1}^{n+1} (-1)^i \xi_i = \sum_{i=1}^{n+1} \xi_i f(\tau_i).$$

Using (19) we arrive at

$$\bar{\varepsilon} = \sum_{i=1}^{n+1} \xi_i f(\tau_i).$$

In the same way we find that

$$\epsilon = \sum_{i=1}^{n+1} x_i f(t_i).$$

Applying (18) and the definition of R we obtain

$$\epsilon = \sum_{i=1}^{n+1} x_i R(t_i), \quad \bar{\epsilon} = \sum_{i=1}^{n+1} \xi_i R(\tau_i). \tag{21}$$

All terms in the two sums (21) have the same sign due to (14) and (20). Therefore the desired conclusion $|\bar{\epsilon}| > |\epsilon|$ follows from (16).

Remark. By passing from the set $\{t_1,\ldots,t_{n+1}\}$ to $\{\tau_1,\ldots,\tau_{n+1}\}$ as described in Lemma (13) above we perform a simplex-like exchange step and obtain an improvement of the lower bound for the obtainable approximation error. We will now show that it is possible to carry out such an exchange as long as $y_1,\ldots,y_n$, ϵ is not an optimal solution of (PA).

(22) Lemma. Use the same notation as in Lemma (13). Assume that there is a $t^* \in [a,b]$ such that

$$R(t^*) > \epsilon. \tag{23}$$

Then there is a set $\{\tau_1,\ldots,\tau_{n+1}\}$ meeting the conditions (14) - (16).

Proof: Since R is continuous on $[\alpha,\beta]$ and $R(t_i) \cdot R(t_{i+1}) < 0$ by (12), it has n zeros $z_1 < z_2 < \ldots < z_n$ such that

$$t_i < z_i < t_{i+1}, \quad i = 1,\ldots,n.$$

First put $\lambda_i = t_i$, $i = 1,2,\ldots,n+1$. Next one of the λ_i will be replaced by t^*. There are the three cases i), ii), iii):

   i) $t < \lambda_1$. If $R(t^*)R(\lambda_1) > 0$ then t^* replaces λ_1; otherwise t^* replaces λ_{n+1}.

  ii) There is an i such that $\lambda_i < t^* < \lambda_{i+1}$. Then t replaces λ_i if $R(t^*) \cdot R(\lambda_i) > 0$; otherwise t^* replaces λ_{i+1}.

 iii) $t^* > \lambda_{n+1}$. Then t^* replaces λ_{n+1} if $R(t^*) \cdot R(\lambda_{n+1}) > 0$; otherwise t^* replaces λ_1.

Put $\tau_i = \lambda_i$, $i = 1,\ldots,n+1$. Then (14) - (16) are satisfied as claimed.

(24) Theorem. $y_1,\ldots,y_{n+1}$ is an optimal solution of (PA) if and only if there are $n+1$ points $\alpha \leq t_1 < t_2 < \ldots < t_{n+1} \leq \beta$ such that (12) is satisfied with $|\epsilon| = y_{n+1}$.

Proof: If (12) is satisfied then optimality follows from Corollary (7). Assume on the other hand that $y_1,\ldots,y_{n+1}$ is an optimal solution of (PA). Since (PA) and (DA) have the same optimal value and (DA) has a solution we may write

$$y_{n+1} = \sum_{i=1}^{q} x_i f(t_i),\tag{25}$$

$$\sum_{i=1}^{q} x_i u_r(t_i) = 0, \quad r = 1,\ldots,n,\tag{26}$$

$$\sum_{i=1}^{q} |x_i| = 1.\tag{27}$$

We need only consider optimal basic solutions of (DA); i.e. we must have $q \leq n+1$. The homogeneous system (26) has a matrix of coefficients with rank $= \min(q,n)$. Hence it has nontrivial solutions only for $q \geq n+1$ and (DA) has therefore no optimal solutions with $q \leq n+1$. Thus $q = n+1$ is the only possibility for optimal basic solutions. Multiplying (26) by y_r and summing over r we find that

$$\sum_{i=1}^{q} x_i \left(\sum_{r=1}^{n} y_r u_r(t_i) \right) = 0.$$

Thus (25) becomes

$$y_{n+1} = \sum_{i=1}^{n+1} x_i \left\{ f(t_i) - \sum_{r=1}^{n} y_r u_r(t_i) \right\}.\tag{28}$$

By Lemma (1) we have $x_i x_{i+1} < 0$. Hence (27) entails

$$\left| \sum_{i=1}^{n+1} (-1)^i x_i \right| = 1.$$

Entering this expression into (28) we arrive at

$$\left| \sum_{i=1}^{n+1} x_i (-1)^i y_{n+1} \right| = \left| \sum_{i=1}^{n+1} x_i \left\{ f(t_i) - \sum_{r=1}^{n} y_r u_r(t_i) \right\} \right|.$$

Since

$$\left| f(t) - \sum_{r=1}^{n} y_r u_r(t) \right| \leq y_{n+1}, \quad t \in [\alpha,\beta],$$

we must conclude that (12) is satisfied for $|\varepsilon| = y_{n+1}$, establishing the desired result.

(29) Remark. Theorem (24) can be used for deriving a nonlinear system of equations to solve (PA) numerically. (12) is a system of $n+1$ equations with the unknowns $y_1,\ldots,y_{n+1}$ and $t_1,\ldots,t_{n+1}$. The missing

n+1 equations are derived by utilizing the fact that the error function
R of Lemma (13) must have a local extremum at t_i, i = 1,...,n+1.

(30) **Theorem**. Let $u_1,...,u_n$ be an extended Chebyshev system of
order two over $[\alpha,\beta]$ and let f be twice continuously differentiable
on the same interval. Then $y_1,...,y_{n+1}$ is the optimal solution of (PA)
if and only if there is a set $\alpha \leq t_1 < t_2 < ... < t_{n+1} \leq \beta$ such that

$$\sum_{r=1}^{n} y_r u_r(t_i) + (-1)^i \epsilon = f(t_i), \quad i = 1,...,n+1, \tag{31}$$

$$(t_i-\alpha)(\beta-t_i) \left\{ \sum_{r=1}^{n} y_r u_r'(t_i) - f'(t_i) \right\} = 0, \quad i = 1,...,n+1, \tag{32}$$

$$y_{n+1} = |\epsilon|. \tag{33}$$

Proof: (31) and (33) follow from Theorem (24). (32) expresses the
fact that the error function has a local extremum at t_i. If $t_i \in (\alpha,\beta)$
then the derivative of the error function must vanish.

The three-phase algorithm is much simpler for (PA) with Chebyshev
systems than in the general case. q is set to n+1 from the outset and
no clustering occurs in Phase 2. In Phase 1 a discretized version of
(PA) is solved by means of an exchange algorithm based on Lemma (13).
For discretized problems convergence is guaranteed by the fact that only
finitely many exchanges can take place and the calculated lower bound
increases in each step. To improve efficiency one generally exchanges
all t_i in each step and seeks to achieve $|R(\tau_i)| > |R(t_i)|$. The
classical Remez algorithm (see e.g. Cheney (1966)) requires that the maximum value of the error function on $[\alpha,\beta]$ be calculated at each step;
but this cannot be achieved by means of a finite number of arithmetic
operations unless further assumptions are made about the structure of the
function f.

Chapter IX
Examples and Applications of Semi-Infinite Programming

In this chapter we shall illustrate how the techniques of semi-infinite programming can be used for the computational treatment of non-trivial problems in a practical context. We remind the reader that important applications have been discussed elsewhere in the book, e.g. in §6, §7, §19 and §20.

§21. A CONTROL PROBLEM WITH DISTRIBUTED PARAMETERS

(1) In this section we shall treat a problem of potential interest for industry. One wants to change the temperature of a metal body by regulating the temperature of its environment. This must be done within a predetermined period of time and the temperature of the environment can only be varied between an upper and a lower value. We shall discuss a simple model problem which is solved in Glashoff and Gustafson (1976). Only one spatial coordinate occurs, but the solution to be presented here could possibly be applied to paralleliepipedic bodies having large extensions in the remaining two dimensions; i.e. when boundary effects can be neglected.

(2) Thus we consider a thin rod which can be heated symmetrically at both ends but is thermally isolated from its surroundings everywhere else. (The rod could be thought of as representing a cut through a plate in its central part. The two large surfaces of the plate are held at the same temperature and heat flows into or out of the interior of the plate. The heat thus propagates perpendicularly to the large surfaces of the plate, not along the surfaces). We select the coordinate system so that the endpoints of the rod are located at -1 and +1. Inside the rod the

temperature is $y(x,t)$ at the point x at the time t, $-1 < x < 1$. We shall study the temperature of the rod for $0 \leq t \leq T$. We assume that the temperature is governed by the heat diffusion equation,

$$y_t(x,t) = y_{xx}(x,t) - q(x)y(x,t), \quad -1 < x < 1, \ 0 \leq t \leq T, \quad (3)$$

where q is a given twice-differentiable function with

$$q(x) = q(-x), \quad 0 \leq x \leq 1. \quad (4)$$

As usual, y_t, y_{xx}, etc. denote partial derivatives. The temperature of the rod is controlled by varying u, the temperature at the two endpoints. The transfer of heat from the rod to the surrounding medium (or vice versa) follows the law

$$\beta y_x(1,t) = u(t) - y(1,t), \quad 0 \leq t \leq T \quad (5)$$

(right endpoint). An analogous equation holds for the left endpoint. Here, β is a positive constant. Combining (3), (4) and (5) we realize that

$$y(-x,t) = y(x,t), \quad -1 < x < 1, \ 0 \leq t \leq T;$$

i.e. y is an even function of x. Therefore we must have

$$y_x(0,t) = 0, \quad 0 \leq t \leq T.$$

We need only consider $y(x,t)$ for $0 \leq x \leq 1$. Let the temperature of the surrounding medium be $u(t)$, $0 \leq t \leq T$, and let $y(x,T)$ be the resulting temperature distribution in the rod at $t = T$ if the temperature at $t = 0$ is

$$y(x,0) = 0, \quad -1 < x < 1.$$

Now let the desired temperature at $t = T$ be $z(x)$ where z is a continuous function with

$$z(x) = z(-x).$$

We now want to compute a function u which is such that $y(x,T)$ approximates $z(x)$ as closely as possible. For physical reasons we must require that u is a bounded function and introduce the constraint

$$0 \leq u(t) \leq 1, \quad 0 \leq t \leq T.$$

For easy reference we collect the equations describing our control problem.

$$y_t(x,t) - y_{xx}(x,t) + q(x)y(x,t) = 0, \quad 0 < x < 1, \ 0 < t \leq T, \quad (6)$$

$$\beta y_x(1,t) + y(1,t) = u(t), \qquad 0 < t \le T, \tag{7}$$

$$y_x(0,t) = 0, \qquad 0 < t \le T, \tag{8}$$

$$y(x,0) = 0, \qquad 0 \le x \le 1, \tag{9}$$

$$0 \le u(t) \le 1, \qquad 0 \le t \le T. \tag{10}$$

If the control function is continuous, one can establish that the system
(6) - (9) has a classical solution $y(x,t)$ where y and its partial
derivatives y_t, y_{xx} are continuous functions for $0 < x \le 1$, $0 < t \le T$.
y is in fact continuous for $0 \le x \le 1$, $0 \le t \le T$. Thus $y(x,T)$ is
continuous for $0 \le x \le 1$. For continuous u, therefore, we can introduce
the linear control operator L through

$$(Lu)(x) = y(x,T), \qquad 0 \le x \le 1,$$

where y is the solution to the problem (6) - (9). We introduce the uniform norm on the space of functions continuous on $[0,1]$ and formulate
our problem as follows:

$$\text{Minimize} \quad ||Lu - z||_\infty \tag{11}$$

when u varies over all continuous functions satisfying (10). It can be
shown that this problem does not in general have an optimal solution.
Hence one extends the class of functions u to get a solvable control
problem. See Glashoff and Gustafson (1976) for details. Here we take a
short cut to arrive more quickly at a computational treatment.

(12) We select an integer $n \ge 1$ and the fixed numbers $t_0, t_1, \ldots, t_n$,
where

$$0 = t_0 < t_1 < \ldots < t_n = T.$$

Next we denote by U the class of piecewise constant functions u satisfying

$$u(t) = \alpha_r, \quad t_{r-1} < t \le t_r, \quad r = 1, \ldots, n. \tag{13}$$

Thus $u \in U$ is uniquely determined by the vector $(\alpha_1, \ldots, \alpha_n)^T$. If
$u \in U$, then Lu can easily be calculated numerically since we start with
$t = 0$ and calculate $y(x,t_1)$, $0 \le x \le 1$ with $u(t) = \alpha_1$. Next we use
the just computed $y(x,t_1)$ as an initial value for y and determine
$y(x,t_2)$, and so on. Therefore we can approximate (11) with the problem

$$\text{Minimize} \quad ||Lu - z||_\infty \quad \text{over all} \quad u \in U. \tag{14}$$

We next introduce the nonnegative basis functions v_r through

$$v_r(t) = \begin{cases} 1, & t_{r-1} < t \leq t_r, \\ 0, & \text{otherwise} \end{cases} \qquad r = 1,\ldots,n. \tag{15}$$

Thus $v_r \in U$, $r = 1,\ldots,n$, and if u is defined by (13) we get

$$u = \sum_{r=1}^{n} \alpha_r v_r. \tag{16}$$

Next we put

$$w_r = Lv_r, \quad r = 1,\ldots,n, \tag{17}$$

giving

$$Lu = \sum_{r=1}^{n} \alpha_r Lv_r = \sum_{r=1}^{n} \alpha_r w_r.$$

Combining (15) and (16) we find that $u \in U$ meets (10) if and only if

$$0 \leq \alpha_r \leq 1, \quad r = 1,\ldots,n. \tag{18}$$

Hence Problem (14) takes the form

$$\text{Minimize } \left\| \sum_{r=1}^{n} \alpha_r w_r - z \right\|_{\infty} . \tag{19}$$

over all $\alpha \in R^n$ subject to (18). We observe that w_r is determined by putting $u = v_r$ in (6) - (9). Problem (18) may now be treated in analogy to the approximation problems in Chapter III. Thus we first recast it into the form

$$\text{Minimize } \alpha_{n+1} \quad \text{when}$$

$$\left| \sum_{r=1}^{n} \alpha_r w_r(x) - z(x) \right| \leq \alpha_{n+1}, \quad 0 \leq x \leq 1 \tag{20}$$

$$0 \leq \alpha_r \leq 1, \quad r = 1,\ldots,n. \tag{21}$$

By replacing (20) and (21) with equivalent simpler inequalities, we obtain

$$\text{Minimize } \alpha_{n+1} \tag{22}$$

over all $\alpha_1,\ldots,\alpha_{n+1}$ subject to the constraints

$$\sum_{r=1}^{n} \alpha_r w_r(x) + \alpha_{n+1} \geq z(x), \quad 0 \leq x \leq 1, \tag{23}$$

$$-\sum_{r=1}^{n} \alpha_r w_r(x) + \alpha_{n+1} \geq -z(x), \quad 0 \leq x \leq 1, \tag{24}$$

$$\alpha_r \geq 0, \quad r = 1,\ldots,n, \tag{25}$$

$$-\alpha_r \geq -1, \quad r = 1,\ldots,n. \tag{26}$$

(22) - (26) is now a linear optimization problem of the type defined in §3. The three-phase algorithm of Chapter VII applies. The fact that the inequality constraints appear in four disjoint groups makes the organization of the calculation somewhat laborious.

(27) We present here a worked example from Glashoff and Gustafson (1976). In (5) - (9), $q(x) = 0$, $0 \leq x \leq 1$, $\beta = 0.1$ and $z(x) = 0.2$ were selected. Several values of T were treated but we discuss here only the case $T = 0.3$. In this Example $w_r(x)$ may be determined in closed form. Let $\mu_1,\ldots$ be the positive roots of the equation

$$\mu \tan \mu = 10.$$

Next determine $A_k \rho_k(x)$ through

$$A_k \rho_k(x) = 2 \sin \mu_k (\mu_k + \cos \mu_k \sin \mu_k)^{-1} \cos \mu_k x.$$

Then $w_r(x)$ is determined from

$$Lu_r(x) = w_r(x) = \sum_{k=1}^{\infty} A_k \mu_k^2 \rho_k(x) \int_0^T u_r(t) \exp(-\mu_k^2(T-t)) dt.$$

$n = 10$ was chosen and t_r were taken equidistant;

$$t_r = 0.03 \cdot r, \quad r = 0,\ldots,10.$$

The problem (22) - (26) was discretized by means of an equidistant grid with 17 points x_i;

$$x_i = (i-1)/17, \quad i = 1,\ldots,17.$$

Then (22) - (26) was replaced by a linear program having 11 variables $\alpha_1,\ldots,\alpha_{11}$ and 54 constraints. The results in Table (32) below emerged. We note that $0 < \alpha_r < 1$ only for $r = 5,8,9,10$.

Next put

$$\bar{f}(x) = \sum_{r=1}^{10} \bar{\alpha}_r w_r(x) - z(x), \tag{28}$$

where $\bar{\alpha}_1,\ldots,\bar{\alpha}_{10}$ is the calculated solution just obtained. The feasibility condition is in this case that

$$|\bar{f}(x)| \leq \bar{\alpha}_{11}, \quad 0 \leq x \leq 1.$$

We find that $\bar{f}$ has local extrema at the 5 gridpoints 0, 0.3125, 0.6250, 0.8750, 1. Thus we assume that (22) - (26) has an optimal solution $\alpha_1,\ldots,\alpha_{11}$ such that the function

$$f = \sum_{r=1}^{10} \alpha_r w_r - z$$

has local extrema at the endpoints 0 and 1 and at 3 interior points which we denote ξ_1,ξ_2,ξ_3. Thus we get the following 8 equations:

$$|f(0)| = \alpha_{11}, \quad |f(1)| = \alpha_{11}, \tag{29}$$

$$|f(\xi_i)| = \alpha_{11}, \quad i = 1,2,3, \tag{30}$$

$$f'(\xi_i) = 0, \quad i = 1,2,3. \tag{31}$$

We use the result of the discretized problem as an approximation of the solution to the linear optimization problem (22) - (26). Thus we put $\alpha_r = \bar{\alpha}_r$ for $r = 1,2,3,4,6,7$ and assume that f and $\bar{f}$ hav the "same shape", i.e. that they have the same number and the same kind of local extrema, thus enabling us to remove the absolute value symbols and select correct signs in (29) and (30). Thus the 8 equations (29) - (31) have the 8 unknowns $\xi_1,\xi_2,\xi_3,\alpha_5,\alpha_8,\alpha_9,\alpha_{10},\alpha_{11}$. The system is solved with the Newton-Raphson method. Lastly, the optimality of the solution hereby obtained is checked by verifying that the complementary slackness conditions with respect to the dual of (22) - (26) are met. For this particular problem it was possible to simplify the general three-phase algorithm due to the special structure of the error curve f. The dual problem appears here only at the verification step. We also see from (32) that the discretization error is rather small.

(32) Table. Calculated results for $T = 0.3$, $n = 10$, 17 equidistant gridpoints in $[0,1]$

Time interval	Index r	Discretized problem	Continuous problem (20)-(24)
0 - 0.12	1,2,3,4	1	1
0.12-0.15	5	0.43638	0.43631
0.15-0.21	6,7	0	0
0.21-0.24	8	0.10835	0.10848
0.24-0.27	9	0.23068	0.23062
0.27-0.30	10	0.19959	0.19959
Optimal value	11	1.060×10^{-4}	1.069×10^{-4}

(33) Exercise. What could happen if the verification of the complementary slackness conditions is left out? Discuss in particular the case when (22) - (26) is discretized with a fine grid!

§22. OPERATOR EQUATIONS OF MONOTONIC TYPE

(1) We shall use the term operator equation for equations having a function u as unknown. Such problems are often formulated as differential equations or integral equations. If the unknown function occurs linearly, then an approximate solution to the operator equation may be calculated by means of reformulating the given problem into an approximation problem of the type discussed in §6 and later in the book.

(2) We illustrate the general idea by discussing the following example. Let K be a continuous function of two variables s and t, defined for $0 \le s \le 1$, $0 \le t \le 1$. Let f and g be two given functions which are defined on [0,1]. We seek a function u satisfying the condition

$$u(0) = 1 \qquad (3)$$

and fulfilling the linear integro-differential equation

$$u'(t) + f(t)u(t) + \int_0^1 K(t,s)u(s)ds = g(t), \qquad 0 \le t \le 1. \qquad (4)$$

Let now $u_1,\ldots,u_n$ be n given functions which are continuously differentiable on [0,1]. We want to approximate the unknown function u with a linear combination

$$u \approx \sum_{r=1}^{n} y_r u_r. \qquad (5)$$

The idea is to enter this approximation into (4) and to minimize the norm of the function

$$\sum_{r=1}^{n} y_r u_r'(t) + f(t) \sum_{r=1}^{n} y_r u_r(t) + \sum_{r=1}^{n} y_r \int_0^1 K(t,s)u_r(s)ds - g(t),$$

$$0 \le t \le 1.$$

Next put

$$v_r(t) = u_r'(t) + f(t)u_r(t) + \int_0^1 K(t,s)u_r(s)ds, \qquad r = 1,\ldots,n. \qquad (6)$$

If we want to approximate g in the uniform norm, we get the task

$$\text{Minimize } y_{n+1} \qquad (7)$$

subject to the constraints

$$\left| \sum_{r=1}^{n} y_r v_r(t) - g(t) \right| \le y_{n+1}, \quad 0 \le t \le 1, \tag{8}$$

$$\sum_{r=1}^{n} y_r u_r(0) = 1. \tag{9}$$

The last relation comes from (3). The problem (7) - (9) can then be reformulated as a linear optimization problem and solved by means of the
general computational schemes of Chapter VII. We notice the similarity
between the approach taken here and that applied in §21. There is no
direct relation between the value y_{n+1} in (7) and the deviation between
a solution u and an approximating linear combination (5). The problem
(7) - (9) is defined and can be solved numerically even if the original
task (3), (4) does not have a solution.

(10) We shall now discuss a general class of operator equations
where the analysis can be carried much further and where the techniques of
semi-infinite programming permit a systematic computational treatment.
We refer here to the so-called operator equations of monotonic type. See
e.g. Collatz (1952). A comprehensive account is given in Protter-Weinberger
(1967) and we refer the interested reader to this text for the mathematical theory of this class of equations. Numerical examples are also given
in Watson (1973). Here we shall illustrate the use of semi-infinite programming on a particular example.

(11) <u>Example</u>. We want to calculate the uniquely determined function u of two variables satisfying

$$\frac{\partial^2 u}{\partial s^2} + \frac{\partial^2 u}{\partial t^2} = 0 \quad \text{on } A = \{(s,t), \quad 0 < s < 1, \quad 0 < t < 1\} \tag{12}$$

$$u(s,t) = f(s,t), \quad s,t \in \text{bd } A, \text{ the boundary of } A. \tag{13}$$

Here, f is a known continuous function.

(12) is a *monotonic* operator equation and one can show the following
result:

$$v(s,t) \le u(s,t) \le w(s,t) \tag{14}$$

whenever v and w are functions of two variables satisfying

$$v(s,t) \le f(s,t) \le w(s,t), \quad (s,t) \in \text{bd } A \tag{15}$$

$$\frac{\partial^2 w}{\partial s^2} + \frac{\partial^2 w}{\partial t^2} \le 0 \le \frac{\partial^2 v}{\partial s^2} + \frac{\partial^2 v}{\partial t^2}, \quad (s,t) \in A. \tag{16}$$

Our goal is to construct functions v and w numerically. Put

$$w(s,t) = \sum_{r=1}^{n} y_r g_r(s,t),$$

where $g_1,\ldots,g_n$ are defined by the expressions $1,s,t,s^2,st,t^2,\ldots,st^{L-1}$, t^L, respectively. Here, L is an integer and

$$n = \frac{1}{2}(L+1)(L+2).$$

$y_1,\ldots,y_n$ are constants to be determined. We get

$$\frac{\partial^2 w}{\partial s^2} + \frac{\partial^2 w}{\partial t^2} = \sum_{r=1}^{n} y_r f_r(s,t)$$

where f_r are calculated from g_r. Thus $f_r(s,t) = 0$, $r = 1,2,3,5$, $f_4(s,t) = f_6(s,t) = 2$, etc. The conditions (15) and (16) imply

$$\sum_{r=1}^{n} y_r g_r(s,t) \geq f(s,t), \quad (s,t) \in bd\ A,$$

$$\sum_{t=1}^{n} y_r f_r(s,t) \leq 0, \quad\quad (s,t) \in A.$$

We want to find a "good" function w, i.e. a function which satisfies the right inequality in (15) and the left inequality in (16) as well as possible. Therefore we

$$\text{minimize } y_{n+1} \tag{17}$$

over all $y_1,\ldots,y_{n+1}$ subject to

$$f(s,t) + y_{n+1} \geq \sum_{r=1}^{n} y_r g_r(s,t) \geq f(s,t), \quad (s,t) \in bd\ A, \tag{18}$$

$$- y_{n+1} \leq \sum_{r=1}^{n} y_r f_r(s,t) \leq 0, \quad\quad (s,t) \in A. \tag{19}$$

The problem (17) - (19) can easily be recast into the following equivalent linear optimization problem:

$$\text{Minimize } y_{n+1} \tag{20}$$

over all $y_1,\ldots,y_n$ subject to

$$\sum_{r=1}^{n} y_r g_r(s,t) \geq f(s,t), \quad (s,t) \in bd\ A, \tag{21}$$

$$y_{n+1} - \sum_{r=1}^{n} y_r g_r(s,t) \geq -f(s,t), \quad (s,t) \in bd\ A, \tag{22}$$

$$-\sum_{r=1}^{n} y_r f_r(s,t) \geq 0, \qquad (s,t) \in A \qquad\qquad (23)$$

$$y_{n+1} + \sum_{r=1}^{n} y_r f_r(s,t) \geq 0, \quad (s,t) \in A. \qquad\qquad (24)$$

This task can be treated with the computational schemes of Chapter VII.
The construction of v is carried out in an analogous manner. (14) can
now be used to calculate pointwise upper and lower bounds for the solution u.

(25) Exercise. Show that the linear optimization problem (20) -
(24) is solvable. Hint: Use Theorem (7) of §11.

§23.   AN AIR POLLUTION ABATEMENT PROBLEM

We shall resume the discussion of the air pollution control problem
of (14) of §3 but now in a more general context. We noted that pollutants
emitted from various sources, e.g. power plants, contaminate the air.
Sooner or later they will reach the ground as fallout. Thus sulfur com-
pounds from power plants burning fossil fuels may damage soils and acidify
waters causing the death of fish. The pollutants are often transported
long distances before they reach the ground. Thus the severe acidifica-
tion of lakes in Scandinavia is caused, to a large extent, by industry in
Great Britain and Central Europe. Similar phenomena have recently been
observed in the U.S. and Canada. In this section we shall develop a model
which incorporates both air pollution and fallout on the ground. The main
difficulty associated with its application to practical problems is the
construction of the transfer functions. Much research is needed in this
area.

(1) Air pollution control model. We use the same notation as in
(14) of §3 but include the fallout as well. Thus we consider an *air
quality control area* S and a *fallout control area* F. S and F need
not coincide. With each source j we associate the transfer functions
V_j and W_j where $V_j(s)$, $s \in S$, is the contribution from source j to
the annual mean concentration in the air of the pollutant considered at
$s \in S$. In the same way, $W_j(t)$, $t \in F$, is the contribution from source j
to the fallout at $t \in F$. Let N sources with strengths g_j be identified. We assume that the combined annual mean pollutant concentration is
given by

$$\sum_{j=1}^{N} g_j V_j(s), \quad s \in S,$$

and that the total fallout is

$$\sum_{j=1}^{N} g_j W_j(t), \quad t \in F.$$

We assume that the contributions add up according to the principle of superposition. The number of sources is fairly large and therefore we combine them into classes as described in (14) of §3. The sources in each class are regulated in the same way. Upon performing this aggregation we write the total concentration in the air at $s \in S$,

$$\sum_{r=1}^{n} v_r(s),$$

and the total fallout at $t \in F$,

$$\sum_{r=1}^{n} w_r(t).$$

Thus source-class r gives rise to the concentration contribution v_r and the fallout contribution w_r.

One reductions strategy is that the emission of class r is reduced by the fraction E_r. Thus $0 \leq E_r \leq 1$, $r = 1,\ldots,n$. Hence the total remaining concentration after regulation is given by

$$\sum_{r=1}^{n} (1-E_r)v_r(s),$$

and the total fallout becomes

$$\sum_{r=1}^{n} (1-E_r)w_r(t).$$

We now require that the remaining concentration and fallout do not surpass given levels g and f. (The standards g and f may be legally imposed.) We also assume that there are upper bounds $e_r \leq 1$, $r = 1,\ldots,n$, for the fractions E_r. (It may not be technically possible to remove the emissions completely.) Therefore the numbers $E_1,\ldots,E_n$ must meet the conditions

$$0 \leq E_r \leq e_r, \quad r = 1,\ldots,n, \tag{2}$$

$$\sum_{r=1}^{n} (1-E_r)v_r(s) \leq g(s), \quad s \in S, \tag{3}$$

$$\sum_{r=1}^{n} (1-E_r)w_r(t) \leq f(t), \quad t \in F. \tag{4}$$

The reduction of emissions entails costs, e.g. for purification of the exhausts or the use of more expensive fuels than otherwise would have been selected. We shall assume here that the costs are defined by the linear function

$$K(E) = \sum_{r=1}^{n} c_r E_r, \tag{5}$$

where $c_1,\ldots,c_n$ are known numbers. The task of minimizing this cost function $K(E)$ subject to the constraints (2) - (4) can be written as a linear optimization problem as follows:

$$\text{Minimize} \quad \sum_{r=1}^{n} c_r E_r \tag{6}$$

subject to the constraints

$$E_r \geq 0, \quad r = 1,\ldots,n, \tag{7}$$

$$-E_r \geq -e_r, \quad r = 1,\ldots,n, \tag{8}$$

$$\sum_{r=1}^{n} E_r v_r(s) \geq -g(s) + \sum_{r=1}^{n} v_r(s), \quad s \in S, \tag{9}$$

$$\sum_{r=1}^{n} E_r w_r(t) \geq -f(t) + \sum_{r=1}^{n} w_r(t), \quad t \in F. \tag{10}$$

The constraint (9) admits the following interpretation: the total reduction must amount at least to the difference between the concentration before reduction and the imposed standard. Conditions (7) and (8) entail that E is restricted to a compact subset of R^n. Thus the problem (6) - (10) is solvable if it is consistent. Consistency means that the standards are met by maximal reduction; i.e.

$$\sum_{r=1}^{n} (1-e_r)v_r(s) \leq g(s), \quad s \in S,$$

and

$$\sum_{r=1}^{n} (1-e_r)w_r(t) \leq f(t), \quad t \in F.$$

The dual of (6) - (10) may be written:

    Maximize

$$-\eta^T e + \sum_{i=1}^{q_1} x_i g_0(s_i) + \sum_{i=1}^{q_2} \xi_i f_0(t_i) \tag{11}$$

over all vectors $\lambda \in R^n$, $\eta \in R^n$, integers q_1, q_2, points $s_i \in S$, $t_i \in T$, and reals x_i, ξ_i, subject to the constraints

$$\lambda_r - \eta_r + \sum_{i=1}^{q_1} x_i v_r(s_i) + \sum_{i=1}^{q_2} \xi_i w_r(t_i) = c_r, \quad r = 1,\dots,n, \tag{12}$$

$$\lambda_r \geq 0, \quad \eta_r \geq 0, \quad r = 1,\dots,n, \quad x_i \geq 0, \quad i = 1,\dots,q_1,$$
$$\xi_i \geq 0, \quad i = 1,\dots,q_2. \tag{13}$$

In (11) we have put

$$g_0(s) = -g(s) + \sum_{r=1}^{n} v_r(s),$$

$$f_0(t) = -f(t) + \sum_{r=1}^{n} w_r(t).$$

The complementary slackness conditions for the dual pair (6) - (10) and (11) - (13) read

$$\lambda^T E = 0, \tag{14}$$

$$\eta^T(e-E) = 0, \tag{15}$$

$$x_i \left\{ \sum_{r=1}^{n} E_r v_r(s_i) - g_0(s_i) \right\} = 0, \quad i = 1,\dots,q_1, \tag{16}$$

$$\xi_i \left\{ \sum_{r=1}^{n} E_r w_r(t_i) - f_0(t_i) \right\} = 0, \quad i = 1,\dots,q_2. \tag{17}$$

The equations (14) - (17) which must be fulfilled for optimal solutions can be analyzed as follows. Since $0 \leq E_r \leq e_r$, we must have $\lambda_r \eta_r = 0$, $r = 1,\dots,n$. If e.g. $E_r = 0$, then (15) entails $\eta_r = 0$ since $e_r - E_r = e_r > 0$. Further, if $x_i > 0$ then

$$\sum_{r=1}^{n} E_r v_r(s_i) = g_0(s_i).$$

Thus the pollutant concentration reaches the highest possible value at s_i. In the same way the level of fallout reaches the standard value at t_i if $\xi_i > 0$. Thus an optimal reductions strategy is associated with q_1 points $s_1,\dots,s_{q_1}$ where the pollutant concentration reaches the highest value and q_2 points $t_1,\dots,t_{q_2}$ where the rate of fallout is

the largest permissible. The positions of these "critical points" are
determined when Problem (6) - (10) is solved numerically.

For this purpose the general three-phase algorithm of Chapter VII may
be used. In Phase i), S in (9) and F in (10) are replaced by the
finite subsets $\{s_1,\ldots,s_N\} \subset S$ and $\{t_1,\ldots,t_L\} \subset T$ and the resulting
linear program is solved. Let the optimal solution then obtained be de-
noted by E*. We find that

$$\sum_{r=1}^{n} E_r^* v_r(s_j) \geq -g(s_j) + \sum_{r=1}^{n} v_r(s_j), \quad j = 1,\ldots,N,$$

$$\sum_{r=1}^{n} E_r^* w_r(t_\ell) \geq -f(t_\ell) + \sum_{r=1}^{n} w_r(t_\ell), \quad \ell = 1,\ldots,L.$$

This means that with the reduction strategy E*, the standards for pollution and fallout are met on the grids. They can hence be violated only
outside the grids and it is possible to derive bounds for how large the
deviation can be. We recall that $0 \leq E_r^* \leq e_r \leq 1$. Hence one can assess
when it is worthwhile to carry out the remaining phases of the algorithm
since the parameters of this problem, e.g. the transfer functions, are
not very accurately determined.

§24.   NONLINEAR SEMI-INFINITE PROGRAMS

In this section we shall illustrate by examples how the computational
scheme of Chapter VII may be extended to problems which are not of the
form of (P) (introduced in §3) or (D) (introduced in §4).

We treat first the class of problems which arise when the preference function of (P) is replaced by a nonlinear *convex* function. Thus
we consider the following task: Let the index set S and the functions
$a_1,\ldots,a_n$ and b be defined as in §3. Suppose that F is convex and
continuously differentiable on R^n. Consider the problem

Minimize F(y)   (1)

over all $y \in R^n$ subject to the constraints

$$\sum_{r=1}^{n} y_r a_r(s) \geq b(s), \quad s \in S. \tag{2}$$

This problem may be reduced to the form of (P). In our further development we shall assume that (2) determines a compact subset of R^n. Here
it will be denoted by K. Then (1), (2) has an optimal solution y*. We

shall now derive relations which can be used for the determination of y^*.
(1) and (2) may be written as:

Minimize y_{n+1} (3a)

subject to $y_{n+1} = F(y)$, $y \in K$. (3b)

Let us now assume that a cube $T = \{x \mid |x_i| \leq F, i = 1,\ldots,n\}$ is known
with $K \in T$. Denote by $\Pi(\eta,\cdot)$, $\eta \in T$, the linear function

$$\Pi(\eta,y) = F(\eta) + \sum_{r=1}^{n} F_r(\eta)(y_r - \eta_r)$$

where $F_r(\eta)$ stands for $\frac{\partial F}{\partial \eta_r}$. Since F is convex we have (c.f. Blum and
Oettli (1975))

$F(y) \geq \Pi(\eta,y)$, $y \in K$,

$F(y) = \sup_{\eta \in T} \Pi(\eta,y) = \Pi(y,y)$.

Hence (3) is equivalent to the problem

Minimize y_{n+1} (4)

subject to

$y_{n+1} \geq \Pi(\eta,y)$, $\eta \in T$, $y \in K$. (5)

(2) gives the condition that $y \in K$. Combining this with (4) and (5) we
finally arrive at the formulation

Minimize y_{n+1} (6)

subject to

$$y_{n+1} - \sum_{r=1}^{n} y_r F_r(\eta) \geq F(\eta) - \sum_{r=1}^{n} \eta_r F_r(\eta), \quad \eta \in T, \tag{7}$$

$$\sum_{r=1}^{n} y_r a_r(s) \geq b(s), \qquad\qquad s \in S. \tag{8}$$

(6) - (8) is a linear optimization problem of the type introduced in §3.
It can be solved by means of the general three-phase algorithm of Chapter
VII. An alternative is to discretize (1), (2) directly.

This generalization may be carried even further. We consider the
following problem:

Program (PG). Let S be a compact subset of R^n and let g be a
function of the two arguments s,y, where $s \in S$ and $y \in R^n$. g is

required to have the properties that the set

$$K = \{y \in R^n \mid g(s,y) \leq 0, \quad s \in S\} \tag{9}$$

is nonempty and compact, and g is twice continuously differentiable on
S × K.

Let G be twice continuously differentiable on R^n. Then Program
(PG) is the task:

$$\text{Minimize} G(y) \tag{10}$$

over all $y \in R^n$ subject to the constraint

$$g(s,y) \leq 0, \quad s \in S. \tag{11}$$

Remark. Program (PG) has a solution since the continuous function
G is to be minimized over the compact subset K of R^n. Program (P) is
a special case of (PG) which occurs if G is linear and $g(s,y) = b(s) - a(s)^T y$. Since G is not assumed to be convex, G may have arbitrarily
many local minima on K, a fact which complicates the numerical treatment.

To a certain extent, a computational scheme for (PG) can be based on
the experiences from (P), even if the implementation on a computer is much
more difficult.

A natural idea is to discretize (PG), i.e. replace S by a finite
grid

$$T = \{s_1,\ldots,s_N\}$$

and approximate (PG) by the task

$$\text{Minimize} G(y) \tag{12}$$

over all $y \in R^n$ subject to the constraints

$$g(s_j,y) \leq 0, \quad j = 1,\ldots,N. \tag{13}$$

Let now L be a positive interpolating operator with nodes $s_1,\ldots,s_N$
(see (8) of §13). We define

$$Lg(s,y) = \sum_{j=1}^{N} w_j(s)g(s_j,y),$$

where the w_j are as in (8) of 13. We next invoke Theorem (10) of §13
to conclude that y satisfies (13) if and only if

$$Lg(s,y) \leq 0, \quad s \in S.$$

Here the discretization (12), (13) of Program (PG) is equivalent to replacing $g(s,y)$ by $Lg(s,y)$ in (11). For convergence results based on this

fact see Gustafson (1981).

We note that the numerical solution of the discretized problem (12), (13) is a nontrivial task along with the verification that the inequalities (13) are consistent and the set K of (9) is nonempty and compact. These matters must be settled analytically, if possible, This is in sharp contrast to Problem (P) where the questions of the consistency and boundedness of the discretized problem are answered as a result of the simplex calculations.

The problem (12), (13) may be treated using the algorithm in Han (1977) or the variant developed by Powell (1978). Thus we may calculate an approximate solution $\bar{y}$ to (PG). It can be used to determine necessary conditions which must be met by optimal solutions to (PG). Our argument parallels that in §16. See also Gustafson (1981) and Watson (1981).

Let y^* be an optimal solution to (PG). Then two cases are possible:

i) $g(s,y^*) \leq 0$, $s \in S$;

ii) There are q points $s \in S$ such that

$$g(s_j, y^*) = 0, \quad i = 1, \ldots, q. \tag{14}$$

In the first case y^* is a solution to the equation

$$\nabla G(y) = 0. \tag{15}$$

But (15) may have other solutions besides y^*. Thus one would need to determine all solutions to (15) and seek out those which meet (11) and render G a minimum.

Next consider Case ii). Put

$$f(s) = g(s, y^*). \tag{16}$$

Then f has a local maximum at s_j, $j = 1, \ldots, q$. Arguing as in §16, we derive conditions apart from (14) which must be met by y^*. Hence y^* may be considered as a solution to the problem of minimizing G subject to (14) and the constraints generated by the fact that f from (16) assumes a maximum at s_j, $j = 1, \ldots, q$.

In the numerical treatment one approximates y^* by $\bar{y}$, a calculated optimal solution to (12), (13), and derives the constraints by replacing the unknown y^* by the calculated $\bar{y}$ in (14) and (16). Hence we arrive at a nonlinear constrained optimization problem. Using Lagrange multipliers as described in Luenberger (1969), Chap. 9, we may derive a nonlinear system of equations which subsequently is solved numerically, e.g. by means of the Newton-Raphson method. Thus we get a direct generalization of the computational procedures described in Chapter VII. An alterna-

tive approach is to apply the algorithms by Han and Powell which were
mentioned earlier.   In either case an independent verification of the
optimality of the calculated solution is called for.

References

Andreasson, D. O. and Watson, G. A.: Linear Chebyshev approximation without Chebyshev sets, BIT 16 (1976), 349-362.

Bartels, R. H.: A penalty linear programming method using reduced-gradient basis-exchange techniques, Linear Algebra and Appl. 29 (1980), 17-32.

Bartels, R. H. and Golub, G. H.: The simplex method of linear programming using LU-decompositions, CACM 12 (1969), 266-268.

Bartels, R. H., Stoer, J. and Zenger, Ch.: A realization of the simplex method based on triangular decompositions. In: "Linear Algebra", J. H. Wilkinson and C. Reinsch (Eds.), Springer-Verlag, Berlin-Heidelberg-New York (1971).

Blum, E. and Oettli, W.: Mathematische Optimierung, Springer-Verlag, Berlin-Heidelberg-New York (1975).

Carasso, C.: L'algorithme d'exchange en optimisation convexe, Thèse, Grenoble (1973).

Charnes, A. and Cooper, W. W.: Management Models and Industrial Applications of Linear Programming, Vols. I,II, J. Wiley & Sons, New York (1961).

Charnes, A., Cooper, W. W. and Henderson, A.: An Introduction to Linear Programming, J. Wiley & Sons, New York (1953).

Charnes, A., Cooper, W. W. and Kortanek, K. O.: Duality, Haar programs and finite sequence spaces, Proc. Nat. Acad. Sci. U.S. 48 (1962), 783-786.

Charnes, A., Cooper, W. W. and Kortanek, K. O.: Semi-infinite programs which have no duality gap, Management Science 12 (1965), 113-121.

Cheney, E. W.: Introduction to Approximation Theory, McGraw-Hill, New York (1966).

Collatz, L.: Aufgaben monotoner Art, Arch. Math. 3 (1952), 366-376.

Collatz, L.: Approximation von Funktionen bei einer oder mehreren Veränderlichen, ZAMM 36 (1956), 198-211.

Collatz, L. and Krabs, W.: Approximationstheorie, B. G. Teubner, Stuttgart, (1973).

Collatz, L. and Wetterling, W.: Optimierungsaufgaben, Zweite Auflage, Springer-Verlag, Berlin-Heidelberg-New York (1971).

Dahlquist, G. and Björck, Å.: Numerical Methods, Prentice-Hall, Englewood Cliffs, New Jersey (1974).

Dantzig, G. B.: Linear Programming & Extensions, Princeton University Press, Princeton, New Jersey (1963).

Duffin, R. J.: Infinite programs. In "Linear Inequalities and Related Systems", H. W. Kuhn and A. W. Tucker (Eds.), Princeton University Press, Princeton, New Jersey (1956), 157-170.

Eckhardt, U.: Theorems on the dimension of convex sets, Linear Algebra and Appl. 12 (1975), 63-76.

Eggleston, H. G.: Convexity, Cambridge University Press, Cambridge (1958).

Fahlander, K.: Computer programs for semi-infinite optimization, TRITA-NA-7312, Department of Numerical Analysis and Computing Science, Royal Institute of Technology, S-10044 Stockholm 70, Sweden.

Gill, P. E. and Murray, W.: A numerically stable form of the simplex algorithm, Linear Algebra and Appl. 7 (1973), 99-138.

Glashoff, K.: Duality theory of semi-infinite programming. In: "Semi-infinite programming", Proc. Int. Colloqu. Bonn. R. Hettich (Ed.), Lecture Notes in Control and Information Sciences 15, Springer-Verlag, Berlin-Heidelberg-New York (1979), 1-16.

Glashoff, K. and Gustafson, S.-Å.: Numerical treatment of a parabolic boundary-value control problem, J. Opt. Th. Appl. 19 (1976), 645-663.

Glashoff, K. and Gustafson, S.-Å.: Einführung in die Lineare Optimierung, Wissenschaftliche Buchgesellschaft, Darmstadt, (1978).

Gorr, W., Gustafson, S.-Å. and Kortanek, K. O.: Optimal control strategies for air quality standards and regulatory policies, Environment and Planning 4 (1972), 183-192.

Gustafson, S.-Å.: On the computational solution of a class of generalized moment problems, SIAM J. Numer. Anal. 7 (1970), 343-357.

Gustafson, S.-Å.: A general three-phase algorithm for nonlinear semi-infinite programming, in Y. P. Brans (Ed.), Operations Research '81, North-Holland Publ. Co., Amsterdam-New York-Oxford (1981), 495-508.

Gustafson, S.-Å. and Kortanek, K. O.: Numerical treatment of a class of semi-infinite programming problems, Nav. Res. Log. Quart. 20 (1973), 477-504.

Gustafson, S.-Å. and Kortanek, K. O.: On the calculation of optimal long-term air pollution abatement strategies for multiple-source areas, Proc. Sixth NATO/CCMS Expert Panel on Air Poll. Model., (1975).

Hadley, G.: Linear Programming, Addison-Wesley Publ. Comp., Reading, Mass., 3rd printing (1964).

Han, S. P.: A globally convergent method for nonlinear programming, J. Opt. Th. Appl. 22 (1977), 297-309.

Hettich, R.: A Newton-method for nonlinear Chebyshev approximation, In: "Approx. Theory", Proc. Int. Colloqu. Bonn, Lecture Notes Math., 556, Springer-Verlag, Berlin-Heidelberg-New York (1976), 222-236.

Hettich, R. (Ed.): "Semi-infinite Programming", Lecture Notes in Control and Information Sciences 15, Springer-Verlag, Berlin-Heidelberg-New York (1979).

Hettich, R. and Zencke, P.: Numerische Methoden der Approximation und semi-infiniten Optimierung, Teubner, Stuttgart, 1982.

Hildenbrand, K. and Hildenbrand, W.: Lineare Ökonomische Modelle, Springer Hochschultext, Berlin-Heidelberg-New York (1975).

Hoffman, K.-H. and Klostermair, A.: A semi-infinite linear programming procedure and applications to approximation problems in optimal control. Approx. Theory II, Proc. Int. Symp. Austin, (1976), 379-389.

Judin, D. B. and Golstein, E. G.: Lineare Optimierung I, Akademie-Verlag, Berlin (1968).

Karlin, S. and Studden, W. J.: Tchebycheff Systems: with Applications in Analysis and Statistics, Interscience Publishers, New York-London-Sydney (1966).

Krabs, W.: Optimierung und Approximation, B. G. Teubner, Stuttgart (1975).

Lorentz, G. G.: Approximation of Functions, Holt, Rinehart and Winston, New York (1966).

Luenberger, D. G.: Optimization by Vector Space Methods, John Wiley & Sons, New York-London-Sydney-Toronto (1969).

Powell, M. J. D.: A fast algorithm for nonlinearly constrained optimization calculations: In: "Numerical Analysis", G. A. Watson (Ed.), Lecture Notes in Mathematics 630, Springer-Verlag, Berlin-Heidelberg-New York (1978).

Protter, M. H. and Weinberger, H. F.: Maximum Principles in Differential Equations, Prentice-Hall, Englewood Cliffs, New Jersey (1967).

Stewart, G. W.: Introduction to Matrix Computation, Academic Press, New York and London (1973).

Stoer, J.: Einführung in die Numerische Mathematik, 2. Auflage. Springer-Verlag, Berlin-Heidelberg-New York (1976).

Watson, G. A.: One-sided approximation and operator equations, J. Inst. Maths. Applic. 12 (1973), 197-208.

Watson, G. A.: On the best linear one-sided Chebyshev approximation, J. Approx. Theory 7 (1973), 48-58.

Watson, G. A.: Globally convergent methods for semi-infinite programming, Department of Mathematics, University of Dundee (1981).

Index

Applied Mathematical Sciences